ENVIRONMENTAL REMEDIATION TECHNOLOGIES, REGULATIONS AND SAFETY

TIMBER INDUSTRY IN THE DAKOTAS

OUTPUT AND USE ASSESSMENTS

ENVIRONMENTAL REMEDIATION TECHNOLOGIES, REGULATIONS AND SAFETY

Additional books in this series can be found on Nova's website under the Series tab.

Additional E-books in this series can be found on Nova's website under the E-book tab.

ENVIRONMENTAL REMEDIATION TECHNOLOGIES, REGULATIONS AND SAFETY

TIMBER INDUSTRY IN THE DAKOTAS

OUTPUT AND USE ASSESSMENTS

JANICE SEGAL
EDITOR

New York

For permission to use material from this book please contact us:
Telephone 631-231-7269; Fax 631-231-8175
Web Site: http://www.novapublishers.com

NOTICE TO THE READER

The Publisher has taken reasonable care in the preparation of this book, but makes no expressed or implied warranty of any kind and assumes no responsibility for any errors or omissions. No liability is assumed for incidental or consequential damages in connection with or arising out of information contained in this book. The Publisher shall not be liable for any special, consequential, or exemplary damages resulting, in whole or in part, from the readers' use of, or reliance upon, this material. Any parts of this book based on government reports are so indicated and copyright is claimed for those parts to the extent applicable to compilations of such works.

Independent verification should be sought for any data, advice or recommendations contained in this book. In addition, no responsibility is assumed by the publisher for any injury and/or damage to persons or property arising from any methods, products, instructions, ideas or otherwise contained in this publication.

This publication is designed to provide accurate and authoritative information with regard to the subject matter covered herein. It is sold with the clear understanding that the Publisher is not engaged in rendering legal or any other professional services. If legal or any other expert assistance is required, the services of a competent person should be sought. FROM A DECLARATION OF PARTICIPANTS JOINTLY ADOPTED BY A COMMITTEE OF THE AMERICAN BAR ASSOCIATION AND A COMMITTEE OF PUBLISHERS.

Additional color graphics may be available in the e-book version of this book.

Library of Congress Cataloging-in-Publication Data

ISBN: 978-1-63117-161-1

Published by Nova Science Publishers, Inc. † New York

CONTENTS

PREFACE

In 2009, there were 13 primary wood-processing mills in North Dakota, 4 more mills than in 2003, and there were 23 active primary wood-processing mills in South Dakota, 2 fewer mills than in 2004. This book focuses on the timber industry in Dakotas and provides an assessment of timber product output and use in 2009.

Chapter 1 – In 2009, there were 13 primary wood-processing mills in North Dakota, 4 more mills than in 2003. These mills processed 68,000 cubic feet of industrial roundwood, of which 66,000 cubic feet was harvested from the State. Another 89,300 cubic feet of the industrial roundwood harvested in North Dakota was sent to primary wood-processing mills in Minnesota. Saw log harvesting accounted for 97 percent of the total harvest. The harvesting of industrial roundwood products resulted in 79,300 cubic feet of logging residues. Primary wood-processing mills generated 987,700 green tons of mill residues; 40 percent of the mill residues were used for domestic fuel. Thirty-three percent of the mill residues generated were not used for other products.

Chapter 2 – In 2009, there were 23 active primary wood-processing mills in South Dakota, 2 fewer mills than in 2004. Industrial roundwood processed by South Dakota mills increased by 4 percent, from 24.9 million cubic feet in 2004 to 26.0 million cubic feet in 2009. More than 80 percent of the industrial roundwood processed by South Dakota mills was harvested from South Dakota. Eighty-five percent of the imported industrial roundwood processed by South Dakota mills came from Wyoming. There was a total of 24.7 million cubic feet of industrial roundwood harvested in South Dakota in 2009, an increase of 13 percent from 2004. Ninety-five percent of the exported industrial roundwood went to mills in Wyoming. Saw logs accounted for 95 percent of the total harvest. The harvesting of industrial roundwood products resulted in 10.4 million cubic feet of harvest residues. Primary wood-processing mills generated 372,000 green tons of mill residues. Nearly 40 percent of the mill residues generated were used by pulp and particleboard industries. Less than 1 percent of the mill residues were not used for other secondary uses.

In: Timber Industry in the Dakotas   ISBN: 978-1-63117-161-1
Editor: Janice Segal   © 2014 Nova Science Publishers, Inc.

Chapter 1

NORTH DAKOTA TIMBER INDUSTRY: AN ASSESSMENT OF TIMBER PRODUCT OUTPUT AND USE*

David E. Haugen and Robert A. Harsel

ABSTRACT

In 2009, there were 13 primary wood-processing mills in North Dakota, 4 more mills than in 2003. These mills processed 68,000 cubic feet of industrial roundwood, of which 66,000 cubic feet was harvested from the State. Another 89,300 cubic feet of the industrial roundwood harvested in North Dakota was sent to primary wood-processing mills in Minnesota. Saw log harvesting accounted for 97 percent of the total harvest.

The harvesting of industrial roundwood products resulted in 79,300 cubic feet of logging residues. Primary wood-processing mills generated 987,700 green tons of mill residues; 40 percent of the mill residues were used for domestic fuel. Thirty-three percent of the mill residues generated were not used for other products.

---

* This is an edited, reformatted and augmented version of the U.S. Forest Service Northern Research Station publication, dated May 2013.

INTRODUCTION

North Dakota's wood products manufacturing industry employs more than 1,900 workers with an output of about $340 million (NAICS 321—wood product manufacturing) (U.S. Census Bureau 2007). This bulletin analyzes recent North Dakota forest industry trends and reports the results of a detailed study of the forest product industry, industrial roundwood production, and associated primary mill wood and bark residue production in North Dakota in 2009. Such detailed information is necessary for long range planning and decisionmaking in wood procurement, economic research, forest resources management, and forest industry development. Likewise, researchers use current forest industry and industrial roundwood information for assessing future research needs and project development.

The 2003 Timber Industrial Assessment for North Dakota (Haugen and Harsel 2005) was used as a primary baseline of comparison for results. As a result of our ongoing efforts to improve the timber product output (TPO) survey's efficiency and reliability, minor changes in previously published data (e.g., Haugen and Harsel 2005) may have occurred because of omissions or correction of errors with the reprocessing of earlier data. Rows and columns of supporting tables in the current report may not sum due to rounding, but data in each table cell are accurately displayed.

Information about the forest land resource of North Dakota is available at the Forest Inventory and Analysis Web site at: http://nrs.fs.fed.us/fia/data-tools/statereports/ND.

STUDY METHODS

This study was a cooperative effort between the North Dakota Forest Service (NDFS) and the Forest Inventory and Analysis (FIA) unit at the Northern Research Station (NRS) of the U.S. Forest Service. The FIA program is responsible for providing forest resource statistics for all ownerships across the United States, including timber products outputs.

Using questionnaires supplied by NRS and designed to determine the size and composition of the State's primary wood-using industry, its use of roundwood, and its generation and disposition of wood residues, NDFS surveyed all known primary wood-using mills. Completed questionnaires were sent to NRS to process and analyze. As part of data processing, all industrial roundwood volumes reported on the questionnaires were converted to standard units of measure using regional conversion factors (Table 1). Timber removals by source of material and harvest residues generated during logging were estimated from standard product volumes using factors developed from logging utilization studies previously conducted by NRS. Data on North Dakota's industrial roundwood receipts were loaded into a regional timber removals database where they were supplemented with data on out-of State uses of North Dakota roundwood to provide a complete assessment of North Dakota's timber product output.

Certain terms used in this report—retained, export, import, production, and receipts— have specialized meanings and relationships unique to the FIA program that surveys timber product output (TPO) (Fig. 1).

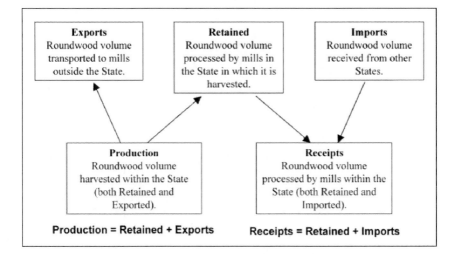

Figure 1. Diagram of the movement of industrial roundwood.

PRIMARY TIMBER INDUSTRY IN NORTH DAKOTA

Industrial Roundwood

- In 2009, North Dakota's primary wood-using industry totaled 13 mills, an increase of 4 mills since 2003 (Table 2, Fig. 2).

Table 1. Conversion factors from reported unit of measure to standard unit of measure[a]

Product (Standard unit of measure)	Reported unit of measure					
	International ¼-inch rule MBF	Doyle scale MBF	Green tons	Standard cords	Thousand pieces	Thousand cubic feet
Saw logs and handles						
(MBF International ¼-inch rule)	1	1.38	0.2174	0.5		0.158
Veneer logs and cooperage						
(MBF International ¼-inch rule)	1	1.14		0.5		0.158
Pulp and composite products, and industrial fuelwood						
(Standard cords)			0.4167	1		0.079
Mine timbers						
(Thousand cubic feet)		0.2322		0.079	6.7	1
Poles						
(Pieces)	20		4.348	10	1,000	0.0079
Posts						
(Thousand pieces)	0.2		0.04167	0.1	1	0.79
Cabin logs, excelsior/shavings, and miscellaneous products						
(Thousand cubic feet)	0.158	0.21804	0.0329193	0.079	7.9	1

a Reported volume times conversion factor = Standard volume.

Table 2. Number of active primary wood-using mills by mill type and survey year, North Dakota

Kind of mill and mill size	Survey Year			
	1993	1998	2003	2009
Sawmills				
Large[a]	1	--	--	--
Small[b]	11	8	9	12
Total	12	8	9	12
Other products[c]	1	1	--	1
All mills	13	9	9	13
Pulp mills	2	2	2	1
Charcoal[d]	60	52	36	15
Handle mills	12	7	10	6
Post and pole mills	14	22	28	23
Other products[e]	44	9	3	11
Total	171	132	113	81
All mills	1,161	681	599	491

[a] Annual lumber production in excess of 5 million board feet.
[b] Annual lumber production less than 1 million board feet.
[c] Includes plants producing mulch, posts, cabin logs, etc.

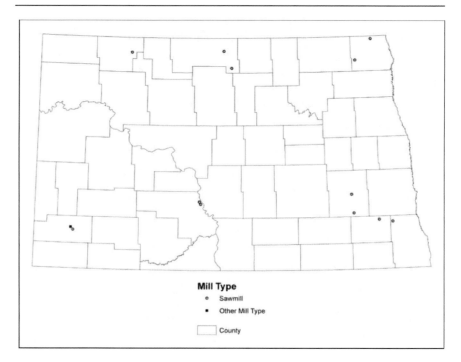

Figure 2. Primary wood-using mills by region, North Dakota, 2009.

- In 2009, the primary wood-using mills in North Dakota processed 68,000 cubic feet of industrial roundwood (Table 3).
- Ninety-eight percent of the industrial roundwood processed by the State's primary wood-using mills was cut from North Dakota forest lands. Minnesota forests supplied the small amount of out-of-State wood used by North Dakota's forest products industry (Table 4).
- Eighty-seven percent of the industrial roundwood processed by North Dakota primary wood-using mills were hardwood species. Cottonwood alone accounted for 83 percent of the total volume processed. Other species of importance to the forest products industry were ponderosa pine, spruce, eastern redcedar, ash, and bur oak.
- Industrial roundwood production decreased by 48 percent, or 295,300 cubic feet in 2003 to 153,200 cubic feet in 2009 (Table 5, Fig. 3).
- Forty-three percent of the 156,300 cubic feet of industrial roundwood harvested in North Dakota was processed in the State (Table 6). Primary wood processors in Minnesota received the remaining 57 percent of the industrial roundwood exported out of state.

Table 3. Industrial roundwood receipts, in thousand cubic feet, by mill type, hardwoods and softwoods, and survey year, North Dakota

Kind of mill	Survey Year				% change from 2003-2009
	1993	1998	2003	2009	
All species					
Saw logs[a]	511.6	59.3	47.9	68.1	42%
Softwoods					
Saw logs[a]	4.9	5.8	0.9	8.8	870%
Hardwoods					
Saw logs[a]	506.7	53.5	47.0	59.3	26%

a Saw logs and other products are combined to avoid disclosure of individual mills.

All table cells without observations are indicated by -- . Table value of 0 indicates the volume rounds to less than 1,000 cubic feet. Columns and rows may not add to their totals due to rounding.

Table 4. Industrial roundwood receipts, in thousand cubic feet, by Forest Inventory Unit, species group, and State of origin, North Dakota, 2009

Species Group	State of Origin		
	Total	Minnesota	North Dacota
Softwoods			
Eastern redcedar	1.3	1.1	0.2
Ponderosa pine	5.3	--	6.2
Spruce	2.2	--	1.2
Softwood total	8.8	1.1	7.6
Hardwoods			

Table 4. (Continued)

Species Group	State of Origin		
	Total	Minnesota	North Dacota
Ash	1.2	--	1.2
Aspen/balsam poplar	0.2	--	0.2
Basswood	0.1	--	0.1
Cottonwood	56.5	--	56.5
Elm	0.4	--	0.4
White oak group	1.0	--	1.0
Hardwood total	59.4	--	59.4
State total	68.1	1.1	67.0

All table cells without observations are indicated by --

Table 5. Industrial roundwood production, in thousand cubic feet, by product, hardwoods and softwoods, and survey year, North Dakota

Product	Survey Year					% change from 2003-2009
	1993	1998	2003	2009		
All species						
Saw logs[a]	516.6	59.3	296.2	156.3		-47%
Softwoods						
Saw logs[a]	9.9	5.7	0.9	7.7		748%
Hardwoods						
Saw logs[a]	506.7	53.5	295.3	153.2		-48%

[a] Saw logs and other products are combined to avoid disclosure of individual mills.

Table 6. Industrial roundwood production, in thousand cubic feet, by Forest Inventory Unit, species group, and State of destination, North Dakota, 2009

Species group	Total	State of Destination	
		Minnesota	North Dacota
Softwoods			
Eastern redcedar	0.2	--	0.2
Ponderosa pine	6.2	--	6.2
Spruce	1.2	--	1.2
Softwood total	7.6	--	7.6
Hardwoods			
Ash	1.8	0.6	1.2
Aspen/balsam poplar	10.3	10.0	0.2
Basswood	0.1	--	0.1
Cottonwood	133.9	77.4	56.5
Elm	0.4	--	0.4
White oak group	2.2	1.2	1.0
Hardwood total	148.6	89.3	59.3
State total	156.2	89.3	66.9

All table cells without observations are indicated by --

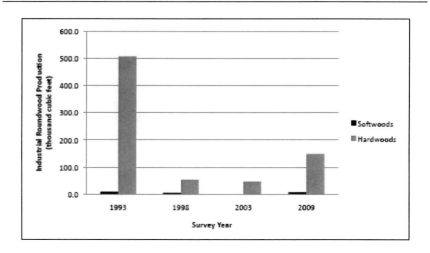

Figure 3. Industrial roundwood production by softwoods and hardwoods, and survey year, North Dakota (Haugen and Harsel 2001, Haugen and Harsel 2005, May and Harsel 1995).

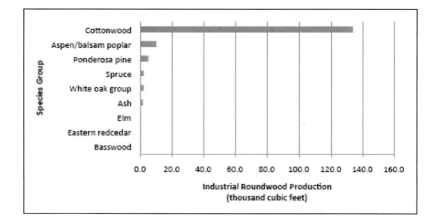

Figure 4. Industrial roundwood production by species group, North Dakota, 2009.

- In 2009, 92 percent or 143,700 cubic feet of industrial roundwood was harvested from the Eastern Forest Inventory Unit (Table 7). Industrial roundwood harvests from the Western Unit were 8 percent (12,500 cubic feet).
- Cottonwood was the most harvested species for industrial roundwood in 2009 (Fig. 4). Other important species harvested were aspen/balsam poplar, ponderosa pine, spruce, bur oak, and ash.

- Sawmills were the largest consumers of North Dakota industrial roundwood produced in 2009. (Table 8, Fig. 5).

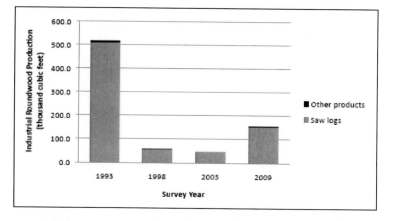

Figure 5. Industrial roundwood production, in thousand cubic feet, by product and survey year, North Dakota (Haugen and Harsel 2001, Haugen and Harsel 2005, May and Harsel 1995).

Saw Logs

- North Dakota sawmill receipts totaled 359,000 board feet in 2009, an increase of 33 percent from 2003 (Table 9). Softwood saw log receipts were estimated at 22,000 board feet, while those of hardwoods equaled 338,000 board feet.
- Cottonwood saw log receipts increased by 49 percent, while aspen/balsam poplar and bur oak saw log receipts declined between the 2003 and 2009 survey.
- Saw log production decreased by 15 percent between 2003 and 2009, from 1 million board feet in 2003 to 859,000 board feet in 2009. Softwood saw log production increased to 16,000 board feet in 2009, while that of hardwoods decreased by 16 percent to 842,000 board feet.
- In 2009, cottonwood accounted for almost 89 percent of the total harvest of saw logs from North Dakota forests. Other important species groups harvested were aspen/balsam poplar, spruce, and bur oak (Fig. 6).
- Residential fuelwood is not included in this report.

Table 7. Industrial roundwood production, in thousand cubic feet, by Forest Inventory Unit, county, and species group, North Dakota, 2009

Forest Inventory Unit and county	All species	Softwoods				Hardwoods						
		Eastern Redcedar	Ponderosa Pine	Spruce	Total Softwoods	Ash	Aspen/balsam poplar	Basswood	Cotton wood	Elm	White oak group	Total hardwoods
Eastern												
Barnes	1.8	--	--	0.2	0.2	0.2	--	--	1.4	--	--	1.6
Bottineau	0.3	--	--	--	--	0.1	0.2	--	--	--	--	0.3
Cass	0.9	--	--	--	--	0.4	--	--	0.3	--	0.2	0.9
Pembina	12.9	--	--	1.0	1.0	--	10.0	--	--	--	1.9	11.9
Ransom	49.3	--	--	--	--	--	--	--	49.3	--	--	49.3
Richland	23.3	0.0	0.7	--	0.7	0.4	--	0.1	22.0	--	0.1	22.6
Traill	55.3	--	--	--	--	0.2	--	--	55.1	--	--	55.3
Unit total	143.8	0.0	0.7	1.2	1.9	1.3	10.2	0.1	128.1	--	2.2	141.9
Western												
Burke	0.6	--	--	--	--	0.1	--	--	0.5	--	--	0.6
Burleigh	3.0	--	--	--	--	0.2	--	--	2.6	0.2	--	3.0
Mortan	3.0	--	--	--	--	0.2	--	--	2.6	0.2	--	3.0
Slope	5.9	0.1	4.6	--	5.7	0.2	--	--	--	--	--	0.2
Unit total	12.5	0.1	4.6	--	5.7	0.7	--	--	5.7	0.4	--	6.8
State total	156.3	0.1	5.3	1.2	7.6	2	10.2	0.1	133.8	0.4	2.2	148.7

All table cells without observations are indicated by -- .

Table 8. Industrial roundwood production by Forest Inventory Unit, species group, and product, North Dakota, 2009

Species group	All units Saw logs MBF[b]	All units Saw logs MCF[a]	Eastern Saw logs MBF[b]	Eastern Saw logs MCF[a]	Western Saw logs MBF[b]	Western Saw logs MCF[a]
Softwoods						
Eastern redcedar	0.9	0.2	0.2	0.0	0.7	0.1
Ponderoda	38.3	6.2	3.8	0.7	34.5	5.5
Spruce	6.4	1.2	6.4	1.2	--	--
Softwood total	16.3	3.0	10.4	1.9	5.9	1.0
Hardwoods						
Ash	10.7	1.8	7.4	1.2	3.4	0.6
Aspen/balsam poplar	55.3	10.3	55.0	10.2	0.3	0
Basswood	0.3	0.1	0.3	0.1	--	--
Cottonwood	760.8	133.9	727.9	128.1	32.9	5.8
Elm	2.0	0.4	--	--	2.0	0.4
White oak group	13.2	2.2	13.2	2.2	--	--
Hardwood total	842.5	148.6	803.9	141.8	38.6	6.8
State total (Unit total)	858.8	151.6	814.3	143.8	44.6	7.8

[a] Thousand cubic feet.

[b] Thousand board feet, International ¼-inch rule.

All table cells without observations are indicated by -- .

Table value of 0 indicates the volume rounds to less than ½ unit of measure. Columns and rows may not add to their totals due to rounding.

Table 9. Saw log receipts and production, in thousand board feet, International 1/4-inch rule, by Forest Inventory Unit and species group, North Dakota, 2003 and 2009

| | All Units | | | | | |
| | Receipts | | | Production | | |
Species group	2003	2009	Percent change	2003	2009	Percent change
Softwoods						
Eastern redcedar	0.3	6.3	2000%	0.3	0.9	200%
Ponderosa pine	0.2	3.8	1800%	0.2	3.8	1800%
Spruce	4.1	11.6	183%	4.1	11.6	183%
Softwood total	4.6	21.7	372%	4.6	16.3	254%
Hardwoods						
Ash	6.6	7.3	11%	6.6	11.1	68%
Aspen/balsam poplar	28.4	1.0	-96%	28.4	55.0	94%
Basswood	2.0	0.3	-85%	2.0	0.3	-85%
Cottonwood	215.0	320.9	49%	952.0	760.8	-20%
Elm	0.3	2.0	567%	0.3	2.0	567%
White oak group	11.7	6.0	-49%	11.7	13.2	13%
Other hardwoods	2.4	--	--	2.4	--	--
Hardwood total	266.4	337.5	27%	1003.4	842.4	-16%
All species	271.0	359.3	33%	1,008.0	858.8	-15%

All table cells without observations are indicated by -- . Table value of 0 indicates the volume rounds to less than 1,000 board feet.

Columns and rows may not add to their totals due to rounding.

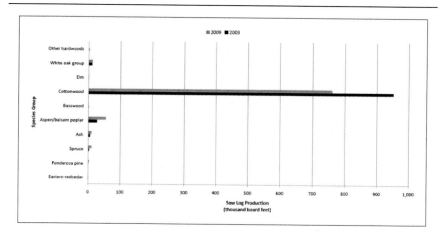

Figure 6. Saw log production by species group, North Dakota, 2003 and 2009.

Timber Removals

During the harvest of industrial roundwood from North Dakota's forests in 2009, 156,200 cubic feet of wood material from growing stock (e.g., sawtimber and pole timber) and non-growing stock (e.g., limb wood, saplings, cull trees, dead trees) was used for primary wood products and another 79,300 cubic feet of wood material from growing stock (e.g., logging residue) and non-growing stock (e.g., logging slash) was left on the ground as harvest residues (Table 10, Fig. 7).

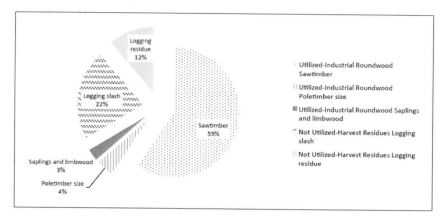

Figure 7. Distribution of timber removals for industrial roundwood by source of material, North Dakota, 2009.

- Growing-stock sources, at 176,300 cubic feet, were the largest component of removals for industrial roundwood production. Eighty-five percent of the growing stock removed was used for products and 15 percent was left as harvest residue. Sawtimber-size trees accounted for 79 percent of the growing-stock volume used for products, and the remainder came from pole-size trees.
- In 2009, 59,300 cubic feet of non-growing-stock wood material was removed in the production of industrial roundwood, but only 11 percent of this material was used for products and the remainder was left on the ground as logging slash. Fifty-eight percent of the non-growing-stock material used for industrial roundwood came from limbs of growing-stock trees and the other 42 percent came from cull trees.
- Ninety-two percent of the total growing-stock material removed from North Dakota's timberland in 2009 came from the Eastern Forest Inventory Unit (Table 11).
- In 2009, 871,000 board feet were removed from North Dakota's sawtimber inventory (Table 12). Cottonwood accounted for 88 percent of the total sawtimber volume removed.
- The harvesting of industrial roundwood products from North Dakota forests in 2009 left 79,300 cubic feet of harvest residues on the ground (Table 13).

Harvest Intensity

- Statewide in 2009, there was an average of 14.0 cubic feet of average annual net growth (gross growth minus mortality) of growing stock per acre on forest land, and an average of 1.8 cubic feet of harvest-related wood removals per acre of forest land in North Dakota. Only two counties had more than 2 cubic feet of total wood material removed per acre of forest land (Fig. 8). (For reference, a cord of roundwood contains about 79 cubic feet of wood.)
- In 2009, there were 741,100 acres of forest land in North Dakota (Haugen 2010). The net volume in live trees on forest land was 679 million cubic feet. The 235,500 cubic feet of total wood material removed due to harvesting (Table 10) was less than 1 percent of the total live volume of trees on forest land in North Dakota.

Table 10. Wood material harvested for industrial roundwood, in thousand cubic feet, by Forest Inventory Unit, source of material, and species group, North Dakota, 2009

Species group	Western										
	Source of material										
	Growing stock				Non-growing stock						
	Used for products				Used for products						
	Sawtimber	Pole timber	Logging residue (not used)	Total growing stock	Lombwood	Saplings	Logging slash (not used)	Total non-growing stock	Total used	Total not used	Total harvested
Softwoods											
Eastern redcedar	0.16	0.01	0.00	0.17	0.01	0.00	0.04	0.04	0.17	0.04	0.21
Ponderosa pine	2.08	0.07	0.06	2.21	0.09	0.00	0.46	0.55	2.24	0.52	2.76
Spruce	5.15	0.06	0.16	5.37	0.05	0.02	1.83	1.90	5.28	1.99	7.27
Softwood total	7.39	0.14	0.23	7.75	0.15	0.02	2.33	2.49	7.69	2.55	10.24
Hardwoods											
Ash	1.80	0.01	0.26	2.07	0.01	0.04	0.46	0.51	1.86	0.72	2.58
Cottonwood	119.91	7.76	25.56	153.23	3.58	2.66	47.78	54.02	133.91	73.35	207.25
Aspen	8.01	2.00	0.31	10.32	0.12	0.07	1.20	1.39	10.19	1.52	11.71
White oak group	2.16	0.01	0.32	2.48	0.01	0.05	0.55	0.61	2.23	0.87	3.09
Basswood	0.05	0.00	0.01	0.07	0.00	0.00	0.02	0.02	0.06	0.03	0.09
Elm	0.32	0.02	0.07	0.40	0.01	0.01	0.13	0.14	0.35	0.19	0.54
Hardwood total	132.25	9.79	26.53	168.58	3.73	2.83	50.14	56.70	148.60	76.68	225.28
State total	139.64	9.93	26.76	176.33	3.88	2.84	52.47	59.19	156.28	79.23	235.52
Softwoods											
Eastern redcedar	0.04	0.00	0.00	0.04	0.00	0.00	0.01	0.01	0.04	0.01	0.05

Table 10. (Continued)

	Western										
	Source of material										
	Growing stock				Non-growing stock						
	Used for products				Used for products						
Species group	Sawtimber	Pole timber	Logging residue (not used)	Total growing stock	Lombwood	Saplings	Logging slash (not used)	Total non-growing stock	Total used	Total not used	Total harvested
Ponderosa pine	1.14	0.04	0.03	1.22	0.05	0.00	0.25	0.30	1.23	0.29	1.52
Spruce	0.64	0.01	0.02	0.67	0.01	0.00	0.23	0.24	0.66	0.25	0.90
Softwood total	1.82	0.05	0.06	1.92	0.06	0.00	0.49	0.55	1.92	0.54	2.47
Hardwoods											
Ash	1.20	0.00	0.18	1.38	0.01	0.03	0.31	0.34	1.24	0.48	1.72
Cottonwood	114.72	7.42	24.46	146.60	3.42	2.55	45.71	51.68	128.12	70.17	198.29
Aspen	8.01	2.00	0.31	10.32	0.12	0.07	1.20	1.39	10.19	1.52	11.71
White oak group	2.16	0.01	0.32	2.48	0.01	0.05	0.55	0.61	2.23	0.87	3.09
Basswood	0.05	0.00	0.01	0.07	0.00	0.00	0.02	0.02	0.06	0.03	0.09
Hardwood total	126.14	9.43	25.27	160.85	3.56	2.69	47.80	54.05	141.83	73.07	214.90
Unit total	127.96	9.48	25.33	162.77	3.62	2.69	48.28	54.60	143.75	73.61	217.37
Softwoods											
Eastern redcedar	0.12	0.00	0.00	0.13	0.01	0.00	0.03	0.03	0.13	0.03	0.16
Ponderosa pine	0.94	0.03	0.03	1.00	0.04	0.00	0.21	0.25	1.01	0.24	1.24
Spruce	4.51	0.05	0.14	4.70	0.05	0.01	1.60	1.66	4.62	1.74	6.36
Softwood total	5.57	0.09	0.17	5.83	0.09	0.01	1.84	1.94	5.76	2.01	7.77
Hardwoods											

	Western										
	Source of material										
	Growing stock				Non-growing stock						
	Used for products				Used for products						
Species group	Sawtimber	Pole timber	Logging residue (not used)	Total growing stock	Lombwood	Saplings	Logging slash (not used)	Total non-growing stock	Total used	Total not used	Total harvested
Ash	0.60	0.00	0.09	0.69	0.00	0.01	0.15	0.17	0.62	0.24	0.87
Cottonwood	5.19	0.34	1.11	6.63	0.15	0.12	2.07	2.34	5.79	3.17	8.97
Elm	0.32	0.02	0.07	0.40	0.01	0.01	0.13	0.14	0.35	0.19	0.54
Hardwood total	6.11	0.36	1.26	7.73	0.17	0.14	2.35	2.65	6.77	3.61	10.38
Unit total	11.67	0.45	1.43	13.55	0.26	0.15	4.18	4.59	12.53	5.62	18.15

[a] Based on factors obtained from regional utilization studies.

All table cells without observations are indicated by -- .

Table value of 0.0 indicates the volume rounds to less than 0.1 thousand cubic feet. Columns and rows may not add to their totals due to rounding.

Table 11. Growing-stock removals from timberland for industrial roundwood, in thousand cubic feet, by Forest Inventory Unit, county, and species group, North Dakota, 2009

Forest Inventory Unit and county	All species	Softwoods				Hardwoods						
		Eastern redcedar	Ponderosa pine	Spruce	Total softwoods	Ash	Aspen/ balsam poplar	Bass-wood	Cotton-wood	Elm	White oak group	Total hardwoods
Eastern												
Barnes	2.0	--	--	0.2	0.2	0.2	--	--	1.6	--	--	1.8
Bottineau	0.2	--	--	--	--	0.1	0.2	--	--	--	--	0.2
Cass	1.1	--	--	--	--	0.5	--	--	0.4	--	0.2	1.1
Pembina	13.3	--	--	1.0	1.0	--	10.1	--	--	--	2.1	12.3
Ransom	56.4	--	--	--	--	--	--	--	56.4	--	--	56.4
Richland	26.5	0.0	0.7	--	0.7	0.4	--	0.1	25.2	--	0.1	25.8
Traill	63.3	--	--	--	--	0.2	--	--	63.0	--	--	63.3
Unit total	162.8	0.0	0.7	1.2	1.9	1.4	10.3	0.1	146.6	--	2.5	160.9
Western												
Burke	0.6	--	--	--	--	0.1	--	--	0.6	--	--	0.6
Burleigh	3.4	--	--	--	--	0.2	--	--	3.0	0.2	--	3.4
Morton	3.4	--	--	--	--	0.2	--	--	3.0	0.2	--	3.4
Slope	6.1	0.1	5.6	--	5.8	0.3	--	--	--	--	--	0.3
Unit total	13.6	0.1	5.6	--	5.8	0.7	--	--	6.6	0.4	--	7.7
State total	176.3	0.2	6.3	1.2	7.7	2.1	10.3	0.1	153.2	0.4	2.5	168.6

All table cells without observations are indicated by -- .

Table value of 0 indicates the volume rounds to less than 1,000 cubic feet. Columns and rows may not add to their totals due to rounding.

Table 12. Sawtimber removals from timberland for industrial roundwood, in thousand board feet, International 1/4-inch rule, by Forest Inventory Unit, county, and species group, North Dakota, 2009

Forest Inventory Unit and county	All species	Softwoods				Hardwoods						
		Eastern redcedar	Ponderosa pine	Spruce	Total softwoods	Ash	Aspen/balsam poplar	Bass-wood	Cotton-wood	Elm	White oak group	Total hardwoods
Eastern												
Barnes	9.9	--	--	0.9	0.9	1.0	--	--	8.0	--	--	9.0
Bottineau	1.0	--	--	--	--	0.3	--	0.7	--	--	--	1.0
Cass	5.5	--	--	--	--	2.4	--	--	1.9	--	1.2	5.5
Pembina	56.3	--	--	5.0	5.0	--	--	40.3	--	--	11.0	51.3
Ransom	280.6	--	--	--	--	--	--	--	280.6	--	--	280.6
Richland	132.3	0.2	3.7	--	3.9	2.2	0.3	--	125.3	--	0.6	128.4
Traill	314.9	--	--	--	--	1.2	--	--	313.6	--	--	314.9
Unit total	800.5	0.2	3.7	5.9	9.8	7.1	0.3	41.0	729.5	--	12.8	790.7
Western												
Burke	3.2	--	--	--	--	0.3	--	--	2.9	--	--	3.2
Burleigh	17.0	--	--	--	--	1.0	--	--	15.0	1.0	--	17.0
Morton	17.0	--	--	--	--	1.0	--	--	15.0	1.0	--	17.0
Slope	33.1	0.6	26.3	4.8	31.7	1.3	--	--	--	--	--	1.3
Unit total	70.3	0.6	26.3	4.8	31.7	3.6	--	--	33.0	2.0	--	38.6
State total	870.8	0.8	30.0	10.8	41.6	10.7	0.3	41.0	762.4	2.0	12.8	829.2

All table cells without observations are indicated by --.
Table value of 0 indicates the volume rounds to less than 1,000 cubic feet. Columns and rows may not add to their totals due to rounding.

Table 13. Harvest residue generated by industrial roundwood harvesting, in thousand cubic feet, by Forest Inventory Unit, county, and species group, North Dakota, 2009

Forest Inventory Unit and county	Softwoods					Hardwoods						
	All species	Eastern redcedar	Ponderosa pine	Spruce	Total softwoods	Ash	Aspen/balsam poplar	Basswood	Cottonwood	Elm	White oak	Total hardwoods
Eastern												
Barnes	0.88	--	--	0.05	0.05	0.07	--	--	0.77	--	--	0.84
Bottineau	0.05	--	--	--	--	0.02	0.03	--	--	--	--	0.05
Cass	0.43	--	--	--	--	0.16	--	--	0.19	--	0.08	0.43
Pembina	2.48	--	--	0.24	0.24	--	1.49	--	--	--	0.74	2.23
Ransom	26.99	--	--	--	--	--	--	--	26.99	--	--	26.99
Richland	12.53	0.01	0.25	--	0.26	0.15	--	0.03	12.05	--	0.04	12.27
Traill	30.26	--	--	--	--	0.08	--	--	30.17	--	--	30.26
Unit total	73.61	0.01	0.25	0.29	0.54	0.48	1.52	0.03	70.17	--	0.87	73.07
Western												
Burke	0.30	--	--	--	--	0.02	--	--	0.28	--	--	0.30
Burleigh	1.61	--	--	--	--	0.07	--	--	1.45	0.10	--	1.61
Morton	1.61	--	--	--	--	0.07	--	--	1.45	0.10	--	1.61
Slope	2.10	0.03	1.74	0.24	2.01	0.09	--	--	--	--	--	0.09
Unit total	5.62	0.03	1.74	0.24	2.01	0.24	--	--	3.17	0.19	--	3.61
State total	79.23	0.04	1.99	0.52	2.55	0.72	1.52	0.03	73.35	0.19	0.87	76.68

All table cells without observations are indicated by --.
Table value of 0 indicates the volume rounds to less than 1,000 cubic feet. Columns and rows may not add to their totals due to rounding.

Primary Mill Residues

- In converting industrial roundwood into products, such as lumber, North Dakota's primary wood-using industries generated 987,700 green tons of wood residue (coarse and fine residues) and bark residue (Table 14).
- Twenty-two percent of the mill residues were in the form of bark residue. Fine wood residue (e.g., sawdust) made up another 28 percent of the total mill residues. Coarse wood residue (e.g., slabs and edgings residue) accounted for the remaining 50 percent (Fig. 9).

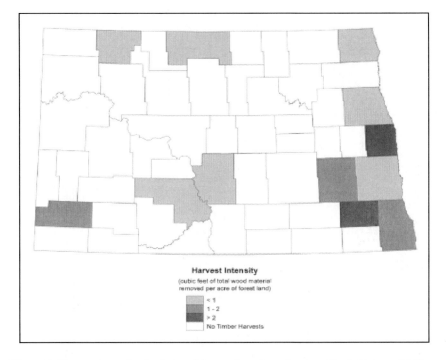

Figure 8. Harvest intensity for industrial roundwood production, North Dakota, 2009.

- Residential fuel, unused, miscellaneous use (e.g., livestock bedding, small dimension, and specialty items), and mulch accounted for 40, 33, 25, and 2 percent, respectively of the end-use of mill residues generated by the primary wood processors in North Dakota (Fig. 10).

Table 14. Disposition of residues produced at primary wood-using mills, in thousand tons, green weight, by Forest Inventory Unit, disposition, residue type, and softwoods and hardwoods, North Dakota, 2009

Forest Inventory Unit and disposition	Total all residues		Total wood residue		Residue type					
					Wood residue				Bark	
					Coarse[a]		Fine[b]			
	Softwood	Hardwood	Softwood	Hardwood	Softwood	Hardwood	Softwood	Hardwood	Softwood	Hardwood
All Units										
Residential fuel	15.8	379.8	15.8	378.0	15.8	375.4	--	2.5	--	1.8
Mulch	15.3	0.4	6.9	--	--	--	6.9	--	8.3	0.4
Miscellaneous[c]	12.4	239.3	10.7	238.9	--	--	10.7	238.9	1.6	0.4
Not used	51.5	273.3	35.3	83.5	22.5	76.9	12.8	6.6	16.2	189.8
Total	94.9	892.8	68.8	700.4	38.3	452.3	30.4	248.0	26.2	192.4
Eastern Unit										
Residential fuel	15.8	325.5	15.8	325.5	15.8	325.5	--	--	--	--
Mulch	--	0.4	--	--	--	--	--	--	--	0.4
Miscellaneous c	8.4	224.0	8.4	223.6	--	--	8.4	223.6	--	0.4
Not used	4.5	250.2	0.5	80.5	--	75.0	0.5	5.5	4.0	169.7
Total	28.7	800.1	24.7	629.7	15.8	400.6	8.9	229.1	4.0	170.4
Western Unit										
Residential fuel	--	54.3	--	52.4	--	49.9	--	2.5	--	1.8
Mulch	15.3	--	6.9	--	--	--	6.9	--	8.3	--
Miscellaneous[c]	3.9	15.3	2.3	15.3	--	--	2.3	15.3	1.6	--
Not used	47.0	23.1	34.9	2.9	22.5	1.8	12.3	1.1	12.2	20.2
Total	66.2	92.7	44.1	70.7	22.5	51.8	21.6	18.9	22.1	22.0

[a] Suitable for chipping such as slabs, edgings, veneer cores, etc.
[b] Not suitable for chipping such as sawdust, veneer clippings etc.
[c] Livestock bedding, small dimension, specialty items, etc. Table may not sum due to rounding.

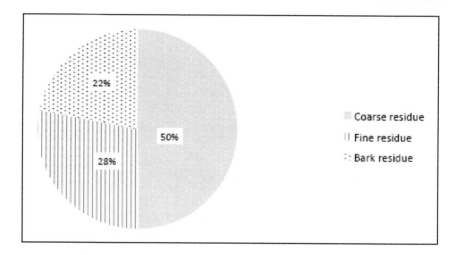

Figure 9. Distribution of residues generated by primary wood-using mills by type of residue, North Dakota, 2009.

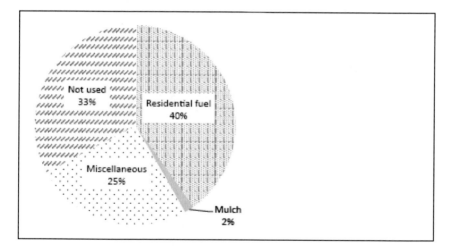

Figure 10. Distribution of residues generated by primary wood-using mills by method of disposal, North Dakota, 2009.

- Seventy-seven percent of the coarse residue was used for domestic fuel. Miscellaneous uses consumed 79 percent of the total fine residue generated, and 92 percent of the bark residue generated went unused.

ACKNOWLEDGMENTS

Special thanks are given to the primary wood-using firms for supplying information for this study and to the North Dakota Forest Service whose cooperation in canvassing survey respondents is greatly appreciated.

Figures 2 and 8 were created by Brian Walters, forester with Forest Inventory and Analysis in St. Paul, MN.

REFERENCES

Haugen, D.E. 2010. North Dakota's forest resources, 2009. Res. Note NRS-83. Newtown Square, PA: U.S. Department of Agriculture, Forest Service, Northern Research Station. 4 p.

Haugen, D. E.; Harsel, R. A. 2005 North Dakota timber industry—an assessment of timber product output and use, 2003. *Resour. Bull.* NC-252. St. Paul, MN:

U.S. Department of Agriculture, Forest Service, North Central Research Station. 18 p.

Haugen, D.E.; Harsel, R.A. 2001 North Dakota timber industry—an assessment of timber product output and use, 1998. Resour. Bull. NC-199. St. Paul, MN: U.S. Department of Agriculture, Forest Service, North Central Research Station. 16 p.

May, D.M.; Harsel, R. A. 1995 North Dakota timber industry—an assessment of timber product output and use, 1993. Resour. Bull. NC-161. St. Paul, MN: U.S. Department of Agriculture, Forest Service, North Central Forest Experiment Station. 14 p.

Miles, P.D. 2011. Forest Inventory EVALIDator Web-application version 4.01 beta.

St. Paul, MN: U.S. Department of Agriculture, Forest Service, Northern Research Station. Available at: http://fiatools.fs.fed.us/Evalidator4/tmattribute.jsp

U. S. Census Bureau. 2007. 2007 Economic census – manufacturing – North Dakota. Available at http://factfinder2.census.gov/faces/nav/jsf/pages/searchresults.xhtml?refresh=t (Accessed May 2011).

In: Timber Industry in the Dakotas
Editor: Janice Segal

ISBN: 978-1-63117-161-1
© 2014 Nova Science Publishers, Inc.

Chapter 2

SOUTH DAKOTA TIMBER INDUSTRY: AN ASSESSMENT OF TIMBER PRODUCT OUTPUT AND USE[*]

Ronald J. Piva and Gregory J. Josten

ABSTRACT

In 2009, there were 23 active primary wood-processing mills in South Dakota, 2 fewer mills than in 2004. Industrial roundwood processed by South Dakota mills increased by 4 percent, from 24.9 million cubic feet in 2004 to 26.0 million cubic feet in 2009. More than 80 percent of the industrial roundwood processed by South Dakota mills was harvested from South Dakota. Eighty-five percent of the imported industrial roundwood processed by South Dakota mills came from Wyoming. There was a total of 24.7 million cubic feet of industrial roundwood harvested in South Dakota in 2009, an increase of 13 percent from 2004. Ninety-five percent of the exported industrial roundwood went to mills in Wyoming. Saw logs accounted for 95 percent of the total harvest. The harvesting of industrial roundwood products resulted in 10.4 million cubic feet of harvest residues. Primary wood-processing mills generated 372,000 green tons of mill residues. Nearly 40 percent of the mill residues generated were used by pulp and particleboard industries.

---

[*] This is an edited, reformatted and augmented version of the U.S. Forest Service publication, dated June 2013.

Less than 1 percent of the mill residues were not used for other secondary uses.

INTRODUCTION

South Dakota's primary wood products manufacturing industry[1] employs 2,470 workers and has a total value of shipments of $590.9 million (U.S. Census Bureau 2007). Given the importance of this industry to the economy of South Dakota, this bulletin analyzes recent forest industry trends and reports the results of a detailed study of forest industry, industrial roundwood production, and associated primary mill wood and bark residue in 2009. Such detailed information is necessary for intelligent planning and decisionmaking in wood procurement, economic research, forest resources management, and forest industry development.

The last published report of timber product output and use in South Dakota was for a 2004 study and is used in this study as a basis for comparison. When new surveys are completed, errors and omissions from previous surveys are corrected. As a result of our ongoing efforts to improve the survey's efficiency and reliability, changes may have been made to the previous survey's data. All comparisons and analysis in this report are based on the reprocessed data from earlier surveys, which may not match earlier published data. Rows and columns of supporting tables may not sum due to rounding, but data in each table cell are accurately displayed.

Information about the forest resources of South Dakota is available at the Forest Inventory and Analysis Web site at: http://nrs.fs.fed.us/fia/data-tools/statereports/SD.

STUDY METHODS

This study was a cooperative effort between the South Dakota Department of Agriculture, Resource Conservation and Forestry Division (SDRCF) and the Forest Inventory and Analysis (FIA) programs at the Northern Research Station (NRS) of the U.S. Forest Service. The FIA program is responsible for providing forest resource statistics for all ownerships across the United States, including timber product outputs.

SDRCF personnel surveyed all known primary wood-using mills, using questionnaires supplied by NRS, to obtain a 100 percent response rate. The questionnaires were designed to determine the size and composition of the State's primary wood-using industry, its use of roundwood, and its generation and disposition of wood residues. Completed questionnaires were sent to the NRS for processing and analysis. As part of data processing, all industrial roundwood volumes reported on the questionnaires were converted to standard units of measure using regional conversion factors (Table 1). Timber removals by source of material and harvest residues generated during logging were estimated from standard product volumes using factors developed from logging utilization studies previously conducted by the NRS. To provide a complete assessment of South Dakota's timber product output, data on the State's industrial roundwood receipts were loaded into a regional timber removals database where they were supplemented with data on out-of-State uses of South Dakota roundwood.

Certain terms used in this report—retained, exports, imports, production, and receipts— have specialized meanings and relationships unique to the FIA program that surveys timber product output (TPO) (Fig. 1). Tables in the appendix relating to saw log volume and sawtimber removal volume are presented in both International 1/4-inch rule and Scribner rule. International 1/4-inch rule is the U.S. Forest Service standard while Scribner rule is the common measure used in South Dakota by forest industries and land management agencies.

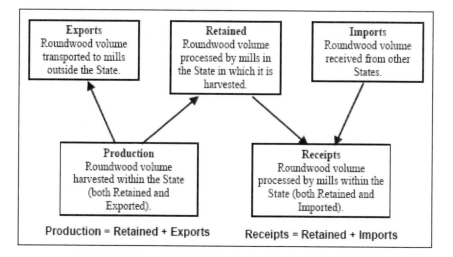

Figure 1.The movement of industrial roundwood.

Table 1. Conversion factors from reported unit of measure to standard unit of measure[a]

Product	International ¼-inch rule	Doyle scale	Scribner scale		Standard	Thousand	Thousand
(Standard unit of measure)	MBF	MBF	MBF	Green tons	cords	pieces	cubic feet
Saw logs and handles (MBF International ¼-inch rule)	1	1.38	1.08	0.2174	0.5	--	0.158
Veneer logs and cooperage (MBF International ¼-inch rule)	1	1.14	1.04		0.5	--	0.158
Pulp and composite products, and industrial fuelwood (Standard cords)	--	--	--	0.4167	1	--	0.079
Poles (Pieces)	20	--	--	4.348	10	1,000	0.0079
Posts (Thousand pieces)	0.2	--	--	0.04167	10.1	1	0.79
Cabin logs, excelsior/shavings, and miscellaneous products (Thousand cubic feet)	0.158	0.21804	0.17604	0.329193	0.079	7.9	1

[a] Reported volume times conversion factor = standard volume. For example, a sawmill reports receiving 100 green tons of industrial roundwood; to convert to MBF International 1/4-inch rule, 100 X 0.2174 = 21.74 MBF International 1/4-inch rule.

PRIMARY TIMBER INDUSTRY
IN SOUTH DAKOTA

Industrial Roundwood

Mill Receipts
- South Dakota's active primary wood-using industry included 13 sawmills, 1 particleboard mill, 3 post and pole mills, and 6 mills that produced cabin logs, excelsior/shavings, or other miscellaneous products (Table 2, Fig. 2).
- The number sawmills in the State decreased from 17 in 2004 to 13 in 2009. Some of the inactive mills indicated that they may resume production in the future.
- Receipts of industrial roundwood at South Dakota primary wood-using mills totaled 26.0 million cubic feet, an increase of 4 percent from the 24.9 million cubic feet received in 2004 (Table 3).
- Eighty-one percent of the industrial roundwood processed by South Dakota's primary wood-using mills was harvested from forests within the State. Wyoming supplied 16 percent of the industrial roundwood consumed by South Dakota mills, with the remainder coming from Montana, Nebraska, and Minnesota (Table 4).
- Nearly 99 percent of the industrial roundwood processed by South Dakota primary wood-using mills was ponderosa pine. Spruce, cottonwood, lodgepole pine, and cedar/ juniper were the next most processed species in 2009.

Industrial Roundwood Production (Harvest)
- Industrial roundwood production increased 13 percent, from 21.8 million cubic feet in 2004 to 24.7 million cubic feet in 2009 (Table 5).
- Ninety-five percent of industrial roundwood harvested in South Dakota was retained for processing by primary wood-using mills in the State. Mills in Wyoming received 95 percent of the industrial roundwood that was exported (Table 6).
- The Western Forest Inventory Unit accounted for nearly 100 percent of total State production. Less that 1 percent of the industrial roundwood harvested in South Dakota came from the Eastern Forest Inventory Unit.

Table 2. Number of active primary wood-using mills by mill type and survey year, South Dakota

Mill type and mill size	Survey Year				
	1993	1999	2004	2009
Sawmills				
Large[a]	3	3	3	2
Medium[b]	3	3	2	2
Small[c]	6	6	12	9
Total	12	12	17	13
Other mills				
Particleboard	1	1	1	1
Post and poles	2	3	3	3
Other mill types[d]	3	2	4	6
Total	6	6	8	10
Total mills	18	18	25	23

[a] Annual lumber production in excess of 5 million board feet.
[b] Annual lumber production from 1 million to 5 million board feet.
[c] Annual lumber production less than 1 million board feet.
[d] Includes excelsior/shavings, cabin log mills, and other miscellaneous products.

Table 3. Industrial roundwood receipts, in million cubic feet, by mill type, survey year, and hardwoods and softwoods, South Dakota

Product	Survey Year				
	1993	1999	2004	2009	2004 – 2009 % change
		ALL SPECIES			
Saw mills	17.9	21.5	23.0	24.8	8%
Post and pole mills	0.0	0.2	0.3	0.9	205%
Other mills[a]	1.0	1.0	1.6	0.3	-84%
Total	18.9	22.6	24.9	26.0	4%
		SOFTWOODS			
Saw mills	17.8	21.2	22.9	24.7	8%
Post and pole mills	0.0	0.2	0.3	0.9	209%
Other mills[a]	1.0	1.0	1.6	0.3	-84%
Total	18.9	22.4	24.8	25.9	4%
		HARDWOODS			
Saw mills	0.1	0.2	0.1	0.0	-45%
Post and pole mills	--	--	0.0	--	--
Other mills[a]	--	--	0.0	0.0	322%
Total	0.1	0.2	0.1	0.0	-46%

All table cells without observations are indicated by -- . Table value of 0.0 indicates the volume rounds to less than 0.1 million cubic feet. Columns and rows may not add to their totals due to rounding.

[a] Includes mills producing excelsior, pulpwood, veneer, cabin logs, etc.

Table 4. Industrial roundwood receipts, in thousand cubic feet, by species group and state of origin, South Dakota, 2009

Species group	Total		State of origin				
		Minnesota	Montana	Nebraska	South Dakota	Wyoming
Softwoods							
Cedar/juniper[a]	18	--	--	9	8	--
Douglas-fir	4	--	4	--	--	0
Lodgepole pine	40	--	33	--	--	6
Ponderosa pine	25,574	--	665	2	20,837	4,069
Other pines	1	1	--	--	--	--
Spruce	276	--	--	--	274	1
Total	25,912	1	702	11	21,120	4,077
Hardwoods							
Ash	2	--	--	--	2	--
Black walnut	3	--	--	--	3	--
Cottonwood	34	--	--	--	34	--
Bur oak	1	--	--	--	0	1
Total	41	--	--	--	40	1
State total	25,952	1	702	11	21,160	4,078

All table cells without observations are indicated by -- . Table value of 0 indicates the volume rounds to less than 1 thousand cubic feet.

Columns and rows may not add to their totals due to rounding.

[a] Includes eastern redcedar and Rocky Mountain juniper.

Table 5. Industrial roundwood production, in million cubic feet, by product, hardwoods and softwoods, and survey year, South Dakota

Product	1993	1999 2004	2009	2004 - 2009% change
		ALL SPECIES		
Saw logs	16.4	20.0	23.6	18%
Cabin logs	--	0.0	0.1	-54%
Posts	0.0	0.2	0.9	189%
Other products[a]	0.9	0.9	0.1	-91%
Total	17.2	21.1	24.7	13%
		SOFTWOODS		
Saw logs	16.2	19.7	23.5	19%
Cabin logs	--	0.0	0.1	-54%
Posts	0.0	0.2	0.9	192%
Other products[a]	0.9	0.9	0.1	-92%
Total	17.0	20.8	24.6	14%
		HARDWOODS		
Saw logs	0.2	0.3	0.1	-58%
Cabin logs	--	--	--	--
Posts	--	--	--	--
Other products[a]	--	--	0.0	--
Total	0.2	0.3	0.1	-58%

Survey Year

All table cells without observations are indicated by -- . Table value of 0.0 indicates the volume rounds to less than 0.1 million cubic feet. Columns and rows may not add to their totals due to rounding.

[a] Includes mills producing excelsior, pulpwood, veneer, poles, and other miscellaneous products.

Table 6. Industrial roundwood production, in thousand cubic feet, by species group, and destination, South Dakota, 2009

Species group	Total		Minnesota	Montana	Destination Nebraska	South Dakota	Wyoming
Softwoods							
Cedar/juniper[a]	8		--	--	--	8	--
Ponderosa pine	24,316		--	163	--	20,837	3,316
Spruce	312		--	--	--	274	38
Total	24,636		--	163	--	21,120	3,354
Hardwoods							
Ash	4		1	--	--	2	--
Black walnut	4		0	--	--	3	--
Cottonwood	44		2	--	8	34	--
Elm	0		0	--	--	--	--
Hard maple	0		0	--	--	--	--
Soft maple	0		0	--	--	--	--
Bur oak	1		0	--	0	0	--
Total	53		4	--	8	40	--
Grand Total	24,689		4	163	8	21,160	3,354

All table cells without observations are indicated by --.

Table value of 0 indicates the volume rounds to less than 1 thousand cubic feet. Columns and rows may not add to their totals due to rounding.

[a] Includes eastern redcedar and Rocky Mountain juniper.

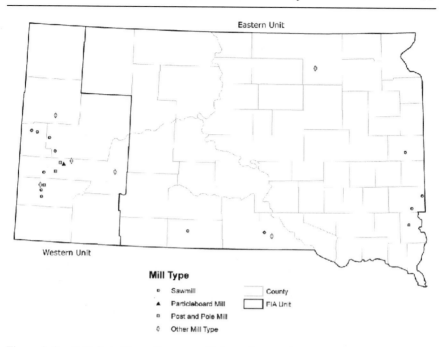

Figure 2. South Dakota Forest Inventory Units and approximate locations of active primary wood-using mills, 2009.

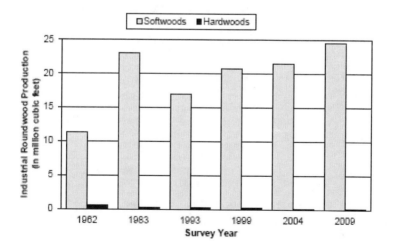

Figure 3. Industrial roundwood production by softwoods and hardwoods, and survey year, South Dakota (Choate and Spencer 1969, Collins and Green 1988, Hackett and Sowers 1996, Piva and Josten 2003, Piva et al. 2006).

- Softwoods, mainly ponderosa pine, accounted for nearly 100 percent of the total industrial roundwood harvested. Spruce, the second most harvested species, accounted for only 1 percent of the total roundwood production (Table 7, Fig. 3).
- The production of saw logs accounted for 95 percent of total industrial roundwood production. Posts and poles were second in production, accounting for 4 percent. The remaining 1 percent of the industrial roundwood products harvested included pulpwood, cabin logs, excelsior/shavings, and other miscellaneous products (Table 8, Fig. 4).

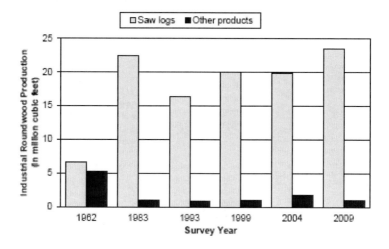

Figure 4. Industrial roundwood production by product and survey year, South Dakota (Choate and Spencer 1969, Collins and Green 1988, Hackett and Sowers 1996, Piva and Josten 2003, Piva et al. 2006).

Saw Logs

The International 1/4-inch rule is the U.S. Forest Service standard unit of measure for volume of saw logs. However, the Scribner rule is a widely applied unit of measure in South Dakota. Therefore, saw log volume will first be presented using the International 1/4-inch rule with the Scribner rule volume following in parentheses.

- Even though the number of active sawmills decreased from 17 mills in 2004 to 13 mills in 2009, total receipts at South Dakota sawmills increased by 8 percent, from 140.9 million board feet (mmbf) (130.4 mmbf Scribner) in 2004 to 152.5 mmbf (141.2 mmbf Scribner) in 2009 (Tables 9 and 9a).

- Saw log production increased by 19 percent between 2004 and 2009, from 121.2 mmbf (112.2 mmbf Scribner) to 144.4 mmbf (133.7 mmbf Scribner). In 2009, saw logs accounted for 95 percent of the total industrial roundwood produced in South Dakota.

- Eighty-five percent of the saw logs harvested in South Dakota were processed by sawmills in the State. Almost 15 percent of the saw logs harvested were sent to sawmills in Wyoming. Sawmills in Minnesota, Montana, and Nebraska (combined) received less than 1 percent of the saw logs harvested.

- Ninety-nine percent of the saw logs harvested in South Dakota in 2009 were ponderosa pine that came from the Western Forest Inventory Unit.

Other Products

- Posts and poles, at 891,000 cubic feet (1.4 million pieces), were the second most harvested product, accounting for 4 percent of the total industrial roundwood harvested from South Dakota's forest in 2009.

- The remaining 1 percent of the industrial roundwood produced in South Dakota was sent to cabin log mills, particleboard mills, excelsior/shavings mills, and other miscellaneous product mills.

- Residential fuelwood is not included in this report.

Timber Removals

- During the harvest of industrial roundwood from South Dakota's forests in 2009, 24.7 million cubic feet of wood material was used for primary wood products and another 10.4 million cubic feet of wood material was left on the ground as harvest residues (Tables 10 and 13, Fig. 5).

- Growing-stock sources (merchantable material), at 26.0 million cubic feet, were the largest component of removals for industrial roundwood production. Ninety-three percent of the growing-stock removed was used for products with 7 percent left as logging residue. Sawtimber-size trees accounted for 97 percent of the growing-stock volume that was used for products.

- Non-growing-stock sources of industrial roundwood amounted to 9.1 million cubic feet of wood material removed. Only 6 percent of this material was used for products, the remainder was left on the ground as logging slash. Nearly half of the non-growingstock material used for industrial roundwood products came from dead trees, mostly ponderosa pine trees killed by mountain pine beetle. The remainder came from cull trees, limbwood, and saplings.

- Nearly 100 percent of the growing-stock material removed from South Dakota's timberland came from the Western Forest Inventory Unit (Table 11). Even though less than 1 percent of the growing-stock volume removed came from the Eastern Forest Inventory Unit, this unit accounted for 98 percent of the hardwood growing-stock removals.

- In 2009, 144.3 million board feet-International 1/4 -inch rule (133.6 million board feet -Scribner rule) of South Dakota's sawtimber inventory was removed from timberland (Tables 12 and 12a).

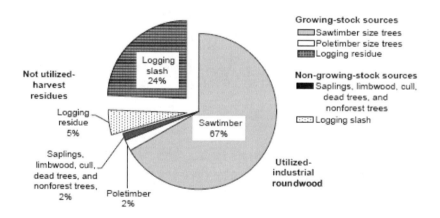

Figure 5. Distribution of timber removals for industrial roundwood by source of material, South Dakota, 2009.

Harvest Intensity

- For 2005 through 2009, FIA reported an average of 20.7 cubic feet of annual net growth of live trees per acre per year on forest land (total annual net growth of live trees divided by forest land area), and 14.8 cubic feet of harvest-related live tree removals per acre per year on forest land (total annual harvest removals of live trees divided by forest land area) (Piva 2010).

- Based on this TPO study for South Dakota, the current removals for the year 2009 averaged 18.6 cubic feet of total harvest removals (removals for product plus harvest residues) per acre of forest land. Only 18 of the 66 total counties in South Dakota reported industrial roundwood removals in 2009, and only 3 counties reported more than 15.0 cubic feet of harvest removals per acre of forest land (Fig. 6). (For reference, a cord of roundwood contains about 80 cubic feet of wood.)

- The Western Forest Inventory Unit averaged 24.1 cubic feet of harvest removals per acre of forest land, and the Eastern Forest Inventory Unit averaged just 0.3 cubic feet of harvest removals per acre of forest land.

- South Dakota's net volume of live trees on forest land in 2009 was 2.3 billion cubic feet (Piva 2010). The 35.1 million cubic feet of total wood material removed due to harvesting (Table 10) represents only 1.5 percent of the total live volume of trees on forest land.

Primary Mill Residues

- In converting industrial roundwood into products, South Dakota's primary wood-using industries generated a combined 371,900 green tons of coarse wood residue (55 percent slabs, edgings, and veneer cores), fine wood residue (20 percent sawdust and veneer clippings), and bark residue (25 percent) (Table 14).

- Thirty-nine percent of mill residues generated were used to make fiber products such as pulp/paper and particleboard. Another 20 percent of the mill residues were used for industrial fuelwood; 19 percent were used to produce wood pellets; 13 percent were used to produce mulch; and 8 percent were used for residential fuelwood and other miscellaneous uses. Only 1 percent of residues generated went unused (Fig. 7).

- The most common disposal method for coarse wood residue and saw dust was for the production of fiber products, with more than 50 percent of the combined total going for this use. Nearly 55 percent of the bark residue was used for industrial fuelwood.

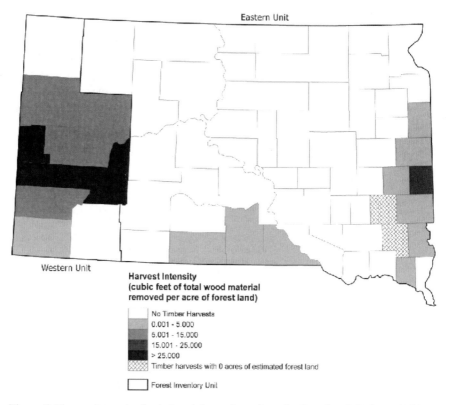

Figure 6. Harvest intensity for industrial roundwood production, South Dakota, 2009.

Table 7. Industrial roundwood production, in thousand cubic feet, by Forest Inventory Unit, county, and species group, South Dakota, 2009

Forest Inventory Unit and county	All species	Softwoods				Species group — Hardwoods							
		Cedar/ junipera	Ponderosa pine	Spruce	Total softwoods	Ash	Black walnut	Cottonwood	Elm	Hard maple	Soft maple	Bur oak	Total hardwoods
Eastern													
Brookings	1	--		--	--	0	1	1	--	--	0	--	1
Clay	8	--		--	--	--	--	8	--	--	--	--	8
Deuel	1	--		--	--	0	0	1	--	--	0	--	1
Gregory	6	5		--	5	0	--	1	--	--	--	0	1
Lake	1	--		--	--	0	--	1	--	--	--	--	1
Lincoln	14	--		--	--	1	1	12	--	0	--	0	14
McCook	1	--		--	--	0	--	1	--	--	--	--	1
Minnehaha	6	--		--	--	1	2	3	0	0	0	--	6
Moody	18	--		--	--	1	1	16	--	--	0	--	18
Todd	33	--	33	33	33	--	--	--	--	--	--	0	0
Tripp	0	--		--	--	--	--	--	--	--	--	0	0
Turner	1	--		--	--	0	--	1	--	--	--	--	1
Total	89	5	33	33	37	4	4	43	0	0	0	1	52
Western													
Butte	96	--	95	1	96	--	--	--	--	--	--	--	--
Custer	3,520	--	3,498	22	3,520	--	--	--	--	--	--	--	--
Fall River	16	--	16	--	16	--	--	--	--	--	--	--	--
Lawrence	8,049	--	7,882	168	8,049	--	--	--	--	--	--	--	--
Meade	512	--	510	1	512	--	--	--	--	--	--	--	--
Pennington	12,407	4	12,282	120	12,406	--	--	1	--	--	--	--	1

Table 7. (Continued)

Forest Inventory Unit and county	All species	Softwoods				Hardwoods							
		Cedar/ junipera	Ponderosa pine	Spruce	Total softwoods	Ash	Black walnut	Cottonwood	Elm	Hard maple	Soft maple	Bur oak	Total hardwoods
Total	24,600	4	24,284	312	24,599	--	--	1	--	--	--	--	1
State total	24,689	8	24,316	312	24,636	4	4	44	0	0	0	1	53

All table cells without observations are indicated by -- . Table value of 0 indicates the volume rounds to less than 1 thousand cubic feet.

Columns and rows may not add to their totals due to rounding.

[a] Includes eastern redcedar and Rocky Mountain juniper.

Table 8. Industrial roundwood production by Forest Inventory Unit, species group, and type of product, South Dakota, 2009

	All Units						
		Type of product					
		Saw logs			Posts and poles		Other products[a]
Species group	All products MCF[b]	International ¼-inch MBF[c]	Scribner MBF[c]	MCF[b]	M pieces[d]	MCFb	MCF[b]
Softwoods							
Cedar/junipere	8	11	10	2	--	--	6
Ponderosa pine	24,316	142,492	131,937	23,227	1,371	891	198
Spruce	312	1,624	1,504	312	--	--	--
Total	24,636	144,127	133,451	23,541	1,371	891	204
Hardwoods							

| | All Units | | | | | | |
| | All products | Saw logs | | | Posts and poles | | Other products[a] |
Species group	MCF[b]	International ¼-inch MBF[c]	Scribner MBF[c]	MCF[b]	M pieces[d]	MCFb	MCF[b]
Ash	4	21	20	4	--	--	--
Black walnut	4	21	19	4	--	--	--
Cottonwood	44	246	228	43	--	--	1
Elm	0	0	0	0	--	--	--
Hard maple	0	2	2	0	--	--	--
Soft maple	0	1	1	0	--	--	--
Bur oak	1	4	3	1	--	--	1
Total	53	296	274	52	--	--	1
State total	24,689	144,423	133,725	23,593	1,371	891	205
Eastern Unit							
Softwoods							
Cedar/junipere	5	11	10	2	--	--	3
Ponderosa pine	33	101	94	17	--	--	16
Total	37	112	104	19	--	--	19
Hardwoods							
Ash	4	21	20	4	--	--	--
Black walnut	4	21	19	4	--	--	--
Cottonwood	43	246	228	43	--	--	--
Elm	0	0	0	0	--	--	--
Hard maple	0	2	2	0	--	--	--
Soft maple	0	1	1	0	--	--	--

Table 8. (Continued)

	All Units						
		Type of product					
Species group	All products	Saw logs			Posts and poles		Other products[a]
	MCF[b]	International ¼-inch MBF[c]	Scribner MBF[c]	MCF[b]	M pieces[d]	MCFb	MCF[b]
Bur oak	1	4	3	1	--	--	--
Total	52	296	274	52	--	--	--
Unit total	89	408	378	70	--	--	19
Softwoods							
Cedar/junipere	4						4
Ponderosa pine	24,284	142,391	131,843	23,211	1,371	891	182
Spruce	312	1,624	1,504	312	--	--	--
Total	24,599	144,015	133,347	23,523	1,371	891	185
Hardwoods							
Cottonwood	1	--	--	--	--	--	1
Total	1	--	--	--	--	--	1
Unit total	24,600	144,015	133,347	23,523	1,371	891	186

All table cells without observations are indicated by -- .

Table value of 0 indicates the volume rounds to less than 1 unit. Columns and rows may not add to their totals due to rounding.

[a] Includes excelsior, pulpwood, cabin logs, and other miscellaneous products.

[b] Thousand cubic feet.

[c] Thousand board feet.

[d] Thousand pieces.

[e] Includes eastern redcedar and Rocky Mountain juniper.

Table 9. Saw log receipts and production, in thousand board feet (International 1A-inch rule), by Forest Inventory Unit and species group, South Dakota, 2004 and 2009

Species group	Receipts			Production		
	2004	2009	Percent Change	2004	2009	Percent Change
Softwoods						
Cedar/juniperaᵃ	30	23	-23%	14	11	-21%
Ponderosa pine	135,613	150,811	11%	115,500	142,492	23%
Spruce	4,796	1,428	-70%	4,992	1,624	-67%
Total	140,439	152,263	8%	120,506	144,127	20%
Hardwoods						
Ash	16	14	-12%	13	21	62%
Aspen	5	--	--	5	--	--
Basswood	0	--	--	0	--	--
Black walnut	12	18	50%	8	21	154%
Cottonwood	349	190	-46%	648	246	-62%
Elm	14	--	--	14	0	-98%
Hackberry	0	--	--	0	--	--
Honeylocust	1	--	--	1	--	--
Hard maple	0	--	--	0	2	717%
Soft maple	--	--	--	--	1	--
Bur oak	12	1	-91%	12	4	-71%
Total	412	223	-46%	702	296	-58%
All species	140,851	152,486	8%	121,208	144,423	19%

All table cells without observations are indicated by -- . Table value of 0 indicates the volume rounds to less than 1 thousand board feet.

Columns and rows may not add to their totals due to rounding.

ᵃ Includes eastern redcedar and Rocky Mountain juniper.

Table 9a. Saw log receipts and production, in thousand board feet (Scribner rule), by Forest Inventory Unit and species group, South Dakota, 2004 and 2009

Species group	Receipts			Production		
	2004	2009	Percent Change	2004	2009	Percent Change
Softwoods						
Cedar/juniper[a]	28	21	-23%	13	10	-21%
Ponderosa pine	125,568	139,640	11%	106,944	131,937	23%
Spruce	4,441	1,322	-70%	4,622	1,504	-67%
Total	130,036	140,984	8%	111,580	133,451	20%
Hardwoods						
Ash	15	13	-12%	12	20	62%
Aspen	5	--	--	5	--	--
Basswood	0	--	--	0	--	--
Black walnut	11	17	50%	8	19	154%
Cottonwood	323	176	-46%	600	228	-62%
Elm	13	--	--	13	0	-98%
Hackberry	0	--	--	0	--	--
Honeylocust	1	--	--	1	--	--
Hard maple	0	--	--	0	2	717%
Soft maple	--	--	--	--	1	--
Bur oak	12	1	-91%	12	3	-71%
Total	381	207	-46%	650	274	-58%
All species	130,417	141,191	8%	112,230	133,725	19%

All table cells without observations are indicated by --. Table value of 0 indicates the volume rounds to less than 1 thousand board feet.

Columns and rows may not add to their totals due to rounding.

[a] Includes eastern redcedar and Rocky Mountain juniper.

Table 10. Wood material harvested for industrial roundwood, in thousand cubic feet, by Forest Inventory Unit, source of material, and species group, South Dakota, 2009

	All Units												
	Source of material										Total wood material used	Total wood material not used	Total wood material harvested
	Growing stock				Used for products		Non-growing stock						
	Used for products		Logging residue (not used)	Total growing stock	Limbwood	Saplings	Cull trees	Dead trees	Logging slash (not used)	Total non-growing stock			
Species group	Sawtimber	Pole-timber											
Softwoods													
Cedar/juniper[b]	5.7	2.5	0.7	8.9	0.1	--	0.0	--	0.6	0.7	8.3	1.3	9.5
Ponderosa pine	23,012.7	803.3	1,801.8	25,617.8	26.2	226.3	--	246.7	8,500.2	8,999.4	24,315.2	10,302.0	34,617.2
Spruce	290.1	9.7	8.8	308.6	12.0	--	0.2	--	64.3	76.5	312.0	73.1	385.1
Total	23,308.4	815.6	1,811.3	25,935.3	38.3	226.3	0.2	246.7	8,565.1	9,076.5	24,635.5	10,376.4	35,011.8
Hardwoods													
Ash	3.4	0.0	0.5	3.9	0.0	--	0.1	--	0.9	1.0	3.5	1.4	4.9
Black walnut	3.2	0.2	0.7	4.1	0.1	--	0.1	--	1.3	1.5	3.6	2.0	5.6
Cottonwood	39.4	2.9	8.4	50.7	1.2	--	0.9	--	15.5	17.5	44.4	23.9	68.2
Elm	0.0	0.0	0.0	0.1	0.0	--	0.0	--	0.0	0.0	0.1	0.0	0.1
Hard maple	0.4	0.0	0.1	0.5	0.0	--	0.0	--	0.1	0.1	0.4	0.2	0.6
Soft maple	0.2	0.0	0.0	0.2	0.0	--	0.0	--	0.1	0.1	0.2	0.1	0.3
Bur oak	0.6	0.0	0.1	0.7	0.0	--	0.0	--	0.2	0.2	0.6	0.2	0.8
Total	47.3	3.2	9.8	60.3	1.3	--	1.0	--	18.0	20.3	52.8	27.8	80.6
State total	23,355.7	818.8	1,821.0	25,995.6	39.6	226.3	1.2	246.7	8,583.1	9,096.9	24,688.3	10,404.1	35,092.4

Table 10. (Continued)

	Eastern Unit												
	Source of material												
	Growing stock				Non-growing stock								
	Used for products		Logging residue (not used)	Total growing stock	Used for products		Cull trees	Dead trees	Logging slash (not used)	Total non-growing stock	Total wood material used	Total wood material not used	Total wood material harvested
Species group	Sawtimber	Pole-timber			Limbwood	Saplings							
Softwoods													
Cedar/juniper[b]	3.4	1.1	0.3	4.8	0.1	--	0.0	--	0.5	0.6	4.6	0.8	5.4
Ponderosa pine	31.8	0.3	2.5	34.7	--	--	-	0.3	11.8	12.1	32.5	14.3	46.8
Total	35.3	1.4	2.8	39.5	0.1	--	0.0	0.3	12.2	12.7	37.1	15.1	52.2
Hardwoods													
Ash	3.4	0.0	0.5	3.9	0.0	--	0.1	--	0.9	1.0	3.5	1.4	4.9
Black walnut	3.2	0.2	0.7	4.1	0.1	--	0.1	--	1.3	1.5	3.6	2.0	5.6
Cottonwood	38.8	2.5	8.3	49.6	1.2	--	0.9	--	15.5	17.5	43.4	23.8	67.1
Elm	0.0	0.0	0.0	0.1	0.0	--	0.0	--	0.0	0.0	0.1	0.0	0.1
Hard maple	0.4	0.0	0.1	0.5	0.0	--	0.0	--	0.1	0.1	0.4	0.2	0.6
Soft maple	0.2	0.0	0.0	0.2	0.0	--	0.0	--	0.1	0.1	0.2	0.1	0.3
Bur oak	0.6	0.0	0.1	0.7	0.0	--	0.0	--	0.2	0.2	0.6	0.2	0.8
Total	46.7	2.8	9.7	59.2	1.3	--	1.0	--	18.0	20.3	51.8	27.7	79.5
State total	82.0	4.1	12.5	98.6	1.4	--	1.0	0.3	30.2	33.0	88.9	42.7	131.6

		Western Unit										
	Source of material											
	Growing stock				**Non-growing stock**							
	Used for products				**Used for products**							
Species group	Sawtimber	Pole-timber	Logging residue (not used)	Total growing stock	Limbwood	Saplings	Cull trees	Dead trees	Logging slash (not used)	Total non-growing stock	Total wood material used	Total wood material not used	Total wood material harvested
Softwoods													
Cedar/juniper[b]	2.2	1.5	0.4	4.1	--	--	--	--	0.1	0.1	3.7	0.5	4.1
Ponderosa pine	22,980.8	803.0	1,799.3	25,583.2	26.2	226.3	--	246.4	8,488.4	8,987.3	24,282.7	10,287.8	34,570.4
Spruce	290.1	9.7	8.8	308.6	12.0	--	0.2	--	64.3	76.5	312.0	73.1	385.1
Total	23,273.1	814.2	1,808.5	25,895.8	38.3	226.3	0.2	246.4	8,552.8	9,063.8	24,598.4	10,361.3	34,959.7
Hardwoods	--	--	--	--	--	--	--	--	--	--	--	--	--
Cottonwood	0.6	0.4	0.1	1.1	--	--	--	--	0.0	0.0	1.0	0.1	1.1
Total	0.6	0.4	0.1	1.1	--	--	--	--	0.0	0.0	1.0	0.1	1.1
State total	23,273.7	814.6	1,808.6	25,896.9	38.3	226.3	0.2	246.4	8,552.8	9,063.9	24,599.4	10,361.4	34,960.8

All table cells without observations are indicated by -- . Table value of 0 indicates the volume rounds to less than 0.1 thousand cubic feet. Columns and rows may not add to their totals due to rounding.

[a] Based on factors obtained from regional utilization studies.

[b] Includes eastern redcedar and Rocky Mountain juniper.

Table 11. Growing-stock removals from timberland for industrial roundwood, in thousand cubic feet, by Forest Inventory Unit, county, and species group, South Dakota, 2009

Forest Inventory Unit and county	All species	Softwoods				Hardwoods							
		Cedar/ junipera	Ponderosa pine	Spruce	Total softwoods	Ash	Black walnut	Cotton-wood	Elm	Hard maple	Soft maple	Bur oak	Total hardwoods
Eastern													
Brookings	2	--	--	--	--	0	1	1	--	--	0	--	2
Clay	9	--	--	--	--	--	--	9	--	--	--	--	9
Deuel	1	--	--	--	--	0	0	1	--	--	0	--	1
Gregory	6	5	--	--	5	0	--	1	--	--	--	0	1
Lake	1	--	--	--	--	0	--	1	--	--	--	--	1
Lincoln	16	--	--	--	--	1	1	14	--	0	--	0	16
McCook	1	--	--	--	--	0	--	1	--	--	--	--	1
Minnehaha	6	--	--	--	--	1	2	3	0	0	0	--	6
Moody	20	--	--	--	--	1	1	19	--	--	0	--	20
Todd	35	--	35	--	35	--	--	--	--	--	--	0	0
Tripp	0	--	--	--	--	--	--	--	--	--	--	0	0
Turner	1	--	--	--	--	0	--	1	--	--	--	--	1
Total	99	5	35	--	39	4	4	50	0	0	0	1	59
Western													
Butte	102	--	101	1	102	--	--	--	--	--	--	--	--

Forest Inventory Unit and county	Softwoods					Hardwoods							
	All species	Cedar/ junipera	Ponderosa pine	Spruce	Total softwoods	Ash	Black walnut	Cotton-wood	Elm	Hard maple	Soft maple	Bur oak	Total hardwoods
Custer	3,597	--	3,575	21	3,597	--	--	--	--	--	--	--	--
Fall River	17	--	17	--	17	--	--	--	--	--	--	--	--
Lawrence	8,559	--	8,393	166	8,559	--	--	--	--	--	--	--	--
Meade	545	--	544	1	545	--	--	--	--	--	--	--	--
Pennington	13,076	4	12,952	119	13,075	--	--	1	--	--	--	--	1
Total	25,897	4	25,583	309	25,896	--	--	1	--	--	--	--	1
State total	25,996	9	25,618	309	25,935	4	4	51	0	0	0	1	60

All table cells without observations are indicated by -- .

Table value of 0 indicates the volume rounds to less than 1 thousand cubic feet. Columns and rows may not add to their totals due to rounding.

[a] Includes eastern redcedar and Rocky Mountain juniper.

Table 12. Sawtimber removals from timberland for industrial roundwood, in thousand board feet (International 1A-inch rule), by Forest Inventory Unit, county, and species group, South Dakota

		Species group								Hardwoods				
Softwoods														
Forest Inventory Unit and county	All species	Cedar/ juniper[a]	Ponderosa pine	Spruce	Total soft-woods	Ash	Black walnut	Cotton-wood	Elm	Hard maple	Soft maple	Bur oak	Total hardwoods
Eastern													
Brookings	8	--	--	--	--	1	3	4	--	--	0	--	8
Clay	44	--	--	--	--	--	--	44	--	--	--	--	44
Deuel	5	--	--	--	--	1	0	4	--	--	0	--	5
Gregory	28	21	--	--	21	2	--	4	--	--	--	1	7
Lake	5	--	--	--	--	1	--	4	--	--	--	--	5
Lincoln	80	--	--	--	--	6	4	69	--	1	--	0	80
McCook	5	--	--	--	--	1	--	4	--	--	--	--	5
Minnehaha	32	--	--	--	--	6	10	15	0	1	0	.	32
Moody	101	--	--	--	--	3	4	94	--	--	1	--	101
Todd	199	--	198	--	198	--	--	--	--	--	--	1	1
Tripp	1	--	--	--	--	--	--	--	--	--	--	1	1
Turner	5	--	--	--	--	1	--	4	--	--	--	--	5
Total	514	21	198	--	219	20	21	247	0	2	1	3	295
Western													
Butte	585	--	580	5	585	--	--	--	--	--	--	--	--
Custer	18,517	--	18,414	104	18,517	--	--	--	--	--	--	--	--
Fall River	98	--	98	--	98	--	--	--	--	--	--	--	--

Forest Inventory Unit and county	Softwoods					Species group			Hardwoods				
	All species	Cedar/juniper[a]	Ponderosa pine	Spruce	Total soft-woods	Ash	Black walnut	Cotton-wood	Elm	Hard maple	Soft maple	Bur oak	Total hardwoods
Lawrence	48,644	--	47,836	808	48,644	--	--	--	--	--	--	--	--
Meade	3,117	--	3,112	5	3,117	--	--	--	--	--	--	--	--
Pennington	72,785	16	72,185	579	72,781	--	--	4	--	--	--	--	4
Total	143,746	16	142,225	1,501	143,742	--	--	4	--	--	--	--	4
State total	144,260	37	142,423	1,501	143,961	20	21	251	0	2	1	3	300

All table cells without observations are indicated by -- .

Table value of 0 indicates the volume rounds to less than 1 thousand board feet. Columns and rows may not add to their totals due to rounding.

[a] Includes eastern redcedar and Rocky Mountain juniper.

Table 12a. Sawtimber removals from timberland for industrial roundwood, in thousand board feet (Scribner rule), by Forest Inventory Unit, county, and species group, South Dakota, 2009

Forest Inventory Unit and county	Softwoods					Species group			Hardwoods				
	All species	Cedar/juniper[a]	Ponderosa pine	Spruce	Total softwoods	Ash	Black walnut	Cotton-wood	Elm	Hard maple	Soft maple	Bur oak	Total hardwoods
Eastern Unit													
Brookings	7	--	--	--	--	0	3	4	--	--	0	--	7
Clay	41	--	--	--	--	--	--	41	--	--	--	--	41

Table 12a. (Continued)

Forest Inventory Unit and county	All species	Softwoods				Species group			Hardwoods				
		Cedar/juniper[a]	Ponderosa pine	Spruce	Total softwoods	Ash	Black walnut	Cotton-wood	Elm	Hard maple	Soft maple	Bur oak	Total hardwoods
Deuel	5	--	--	--	--	1	0	4	--	--	0	--	5
Gregory	26	19	--	--	19	1	--	4	--	--	--	1	6
Lake	5	--	--	--	--	1	--	4	--	--	--	--	5
Lincoln	74	--	--	--	--	5	3	64	--	1	--	0	74
McCook	5	--	--	--	--	1	--	4	--	--	--	--	5
Minnehaha	30	--	--	--	--	5	9	14	0	1	0	--	30
Moody	94	--	--	--	--	3	3	87	--	--	0	--	94
Todd	184	--	183	--	183	--	--	--	--	--	--	1	1
Tripp	1	--	--	--	--	--	--	--	--	--	--	1	1
Turner	5	--	--	--	--	1	--	4	--	--	--	--	5
Total	476	19	183	--	203	19	19	229	0	2	1	3	273
Western Unit													
Butte	541	--	537	5	541	--	--	--	--	--	--	--	--
Custer	17,146	--	17,050	96	17,146	--	--	--	--	--	--	--	--
Fall River	91	--	91	--	91	--	--	--	--	--	--	--	--
Lawrence	45,041	--	44,293	748	45,041	--	--	--	--	--	--	--	--
Meade	2,886	--	2,882	5	2,886	--	--	--	--	--	--	--	--
Pennington	67,394	15	66,838	537	67,390	--	--	4	--	--	--	--	4

	Species group												
Softwoods						Hardwoods							
Forest Inventory Unit and county	All species	Cedar/ juniper[a]	Ponderosa pine	Spruce	Total softwoods	Ash	Black walnut	Cotton-wood	Elm	Hard maple	Soft maple	Bur oak	Total hardwoods
Total	133,098	15	131,690	1,390	133,094	--	--	4	--	--	--	--	4
State total	133,574	34	131,873	1,390	133,297	19	19	233	0	2	1	3	277

All table cells without observations are indicated by --.

Table value of 0 indicates the volume rounds to less than 1 thousand board feet. Columns and rows may not add to their totals due to rounding.

[a] Includes eastern redcedar and Rocky Mountain juniper.

Table 13. Harvest residue generated by industrial roundwood harvesting, in thousand cubic feet, by Forest Inventory Unit, county, and species group, South Dakota, 2009

	Species group												
Softwoods						Hardwoods							
Forest Inventory Unit and county	All species	Cedar/ juniper[a]	Ponderosa pine	Spruce	Total softwoods	Ash	Black walnut	Cotton-wood	Elm	Hard maple	Soft maple	Bur oak	Total hardwoods
Eastern Brookings	1	--	--	--	--	0	0	0	--	--	0	--	1
Clay	4	--	--	--	--	--	--	4	--	--	--	--	4
Deuel	0	--	--	--	--	0	0	0	--	--	0	--	0
Gregory	1	1	--	--	1	0	--	0	--	--	--	0	1
Lake	0	--	--	--	--	0	--	0	--	--	--	--	0

Table 13. (Continued)

| Forest Inventory | Species group | | | | | | | | |
| | All | Softwoods | | | Hardwoods | | | | |
		Cedar/[a]	Ponderosa	Total	Black	Cotton-	Hard	Soft	Total
Lincoln	7	--	--	--	0	7	0	0	7
McCook	0	--	--	--	--	0	--	--	0
Minnehaha[a]	3	--	--	0	1	1	0	0	3
Moody	10	--	--	0	0	9	--	0	10
Todd	14	--	14	14	--	--	--	--	0
Tripp	0	--	--	--	--	--	--	--	0
Turner	0	--	--	0	--	0	--	--	0
Total	43	1	14	15	2	24	0	0	28
Western									
Butte	42	0	42	42	--	--	--	--	--
Custer	1,357	5	1,352	1,357	--	--	--	--	--
Fall River	7	--	7	7	--	--	--	--	--
Lawrence	3,484	39	3,444	3,484	--	--	--	--	--
Meade	224	0	224	224	--	--	--	--	--
Pennington	5,247	28	5,219	5,247	--	--	0	--	0
Total	10,361	73	10,288	10,361	--	--	0	--	0
State total	10,404	74	10,302	10,376	2	24	0	0	28

All table cells without observations are indicated by -- . Table value of 0 indicates the volume rounds to less than 1 thousand cubic feet.

Columns and rows may not add to their totals due to rounding.

[a] Includes eastern redcedar and Rocky Mountain juniper.

Table 14. Disposition of residues produced at primary wood-using mills, in thousand green tons, by Forest Inventory Unit, disposition, residue type, and softwoods and hardwoods, South Dakota, 2009

Forest Inventory Unit and disposition	Total all residues		Total wood residue		Residue type					
					Wood residue				Bark	
					Coarse[a]		Fine[b]			
	Softwood	Hardwood	Softwood	Hardwood	Softwood	Hardwood	Softwood	Hardwood	Softwood	Hardwood
All Units										
Fiber products	143.18	--	143.18	--	104.90	--	38.28	--	--	--
Industrial fuelwood	75.58	0.22	26.15	0.16	19.57	0.15	6.58	0.00	49.43	0.07
Residential fuelwood	2.31	0.17	1.90	0.15	1.90	0.15	--	--	0.41	0.03
Wood pellets	71.14	--	71.14	--	53.25	--	17.90	--	--	--
Mulch	50.13	0.03	9.55	0.03	5.69	--	3.86	0.03	40.59	0.00
Miscellaneous uses[c]	26.92	0.09	26.92	0.09	19.63	0.00	7.30	0.09	0.00	0.00
Not used	2.04	0.06	1.40	0.02	0.71	0.01	0.69	0.01	0.64	0.04
Total	371.31	0.57	280.24	0.44	205.64	0.31	74.60	0.13	91.07	0.13
Eastern Unit										
Industrial fuelwood	0.04	0.22	0.03	0.16	0.02	0.15	0.01	0.00	0.01	0.07
Residential fuelwood	0.13	0.17	0.13	0.15	0.13	0.15	--	0.15	--	0.02
Mulch	0.03	0.02	0.02	0.02	0.02	--	--	0.02	0.01	--
Miscellaneous uses[c]	0.05	0.09	0.05	0.09	0.05	--	0.00	0.09	0.00	--
Not used	0.27	0.05	0.15	0.01	0.00	--	0.14	0.01	0.12	0.04
Total	0.51	0.55	0.38	0.42	0.23	0.30	0.15	0.12	0.13	0.13
Western Unit										
Fiber products	143.18	--	143.18	--	104.90	--	38.28	--	--	--
Industrial fuelwood	75.54	--	26.12	--	19.54	--	6.57	--	49.43	--
Residential fuelwood	2.18	0.00	1.77	0.00	1.77	0.00	--	0.00	0.41	0.00

Table 14. (Continued)

Forest Inventory	Total all residues		Total wood residue		Residue type					
					Wood residue					
					Coarse[a]		Fine[b]		Bark	
Unit and disposition	Softwood	Hardwood	Softwood	Hardwood	Softwood	Hardwood	Softwood	Hardwood	Softwood	Hardwood
Wood pellets	71.14	--	71.14	--	53.25	--	17.90	--	--	--
Mulch	50.11	0.00	9.53	0.00	5.67	--	3.86	0.00	40.58	0.00
Miscellaneous uses[c]	26.88	0.00	26.88	0.00	19.58	0.00	7.30	0.00	0.00	0.00
Not used	1.77	0.02	1.25	0.01	0.71	0.01	0.55	0.01	0.52	0.00
Total	370.80	0.02	279.86	0.02	205.41	0.01	74.45	0.01	90.93	0.01

All table cells without observations are indicated by -- . Table value of 0.00 indicates the volume rounds to less than 5 green tons.

Columns and rows may not add to their totals due to rounding.

[a] Suitable for chipping such as slabs, edgings, veneer cores, etc.

[b] Not suitable for chipping such as sawdust, veneer clippings, etc.

[c] Livestock bedding, small dimension, and specialty products.

ACKNOWLEDGMENTS

Special thanks are given to the primary wood-using firms for supplying information for this study and to the South Dakota Department of Agriculture, Resource Conservation and Forestry Division, whose cooperation in canvassing survey respondents is greatly appreciated.

Figures 2 and 6 were created by Brian Walters, forester with Forest Inventory and Analysis in St. Paul, MN.

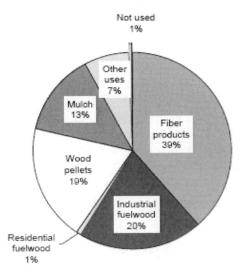

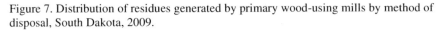

Figure 7. Distribution of residues generated by primary wood-using mills by method of disposal, South Dakota, 2009.

APPENDIX. DEFINITION OF TERMS

Board foot. Unit of measure applied to roundwood. It relates to lumber that is 1 foot long, 1 foot wide, and 1 inch thick (or its equivalent).

Bolt. A short log no more than 8 feet long, to be sawn for lumber, peeled or sliced for veneer, shaved for excelsior, or converted into shingles, cooperage stock, dimension stock, blocks, blanks, or other products.

Central stem. hT e portion of a tree between a 1-foot stump and the minimum 4.0-inch top diameter outside bark, or point where the central stem breaks into limbs.

Coarse mill residue. Wood residue suitable for chipping such as slabs, edgings, and veneer cores.

Commercial species. Tree species presently or prospectively suitable for industrial wood products. (Note: Excludes species of typically small size, poor form, or inferior quality such as hophornbeam, Osage-orange, and redbud.)

Cull removals. Net volume of rough and rotten trees plus the net volume in sections of the central stem of growing-stock trees that do not meet regional merchantability standards but are harvested for industrial roundwood products.

Diameter at breast height (d.b.h.). The outside bark diameter at 4.5 feet above the forest floor on the uphill side of the tree. For determining breast height, the forest floor includes the duff layer that may be present, but does not include unincorporated woody debris that may rise above the ground line.

Doyle rule. A simple log rule or formula for estimating the board-foot volume of logs based on a 4-inch slabbing allowance to square the log. hT is rule is used in the Eastern and Southern United States.

Exports. The volume of roundwood utilized by mills outside the state where the timber was harvested.

Fine mill residue. Wood residue not suitable for chipping, such as sawdust and veneer clippings.

Forest land. Land at least 10 percent stocked with trees of any size, or formerly having had such tree cover, and not currently developed for nonforest use. (Note: Stocking is measured by comparing specified standards with basal area and/or number of trees, age or size, and spacing.) The minimum area for classification of forest land is 1 acre. Roadside, streamside, and shelterbelt strips of timber must have a crown width of at least 120 feet to qualify as forest land. Unimproved roads and trails, streams or other bodies of water, or clearings in forest areas shall be classified as forest if less than 120 feet wide.

Growing-stock removals. The growing-stock volume removed from timberland by harvesting industrial roundwood products. (Note: Includes sawtimber removals, poletimber removals, and logging residues.)

Growing-stock tree. A live timberland tree of commercial species that meets specified standards of size, quality, and merchantability. (Note: Excludes rough, rotten, and dead trees.)

Growing-stock volume. Net volume of growing-stock trees 5.0 inches d.b.h. and larger, from 1 foot above the ground to a minimum 4.0-inch top diameter outside bark of the central stem or to the point where the central stem breaks into limbs.

Hardwoods. Dicotyledonous trees, usually broad-leaved and deciduous.

Harvest residues. The total net volume of unused portions of trees cut or killed by logging. (Note: Includes both logging residues and logging slash.)

Industrial fuelwood. A roundwood product, with or without bark, used to generate energy at manufacturing facilities and schools, correctional institutions, or electric generating plants.

Imports. hT e volume of roundwood delivered to a mill or group of mills in a specific state but harvested outside that state.

Industrial roundwood exports. hT e quantity of industrial roundwood harvested in a geographical area and transported to other geographical areas.

Industrial roundwood imports. The quantity of industrial roundwood received from other geographical areas.

Industrial roundwood products. Saw logs, pulpwood, veneer logs, poles, commercial posts, pilings, cooperage logs, particleboard bolts, shaving bolts, lath bolts, charcoal bolts, and chips from roundwood used for pulp or board products.

Industrial roundwood production. The quantity of industrial roundwood harvested in a geographic area plus all industrial roundwood exported to other geographical areas.

Industrial roundwood receipts. The quantity of industrial roundwood received by commercial mills in a geographic area plus all industrial roundwood imported from other geographical areas.

Industrial roundwood retained. The quantity of industrial roundwood harvested from and processed by commercial mills within the same geographical area.

International 1/4-inch rule. A log rule or formula for estimating the board-foot volume of logs, allowing 1/2 inch of taper for each 4-foot length and assuming 1/4 inch of kerf. hT is rule is used as the U.S. Forest Service standard log rule in the Eastern United States.

Limbwood removals. Net volume of all portions of a tree other than the central stem (including forks, large limbs, tops, and stumps) harvested for industrial roundwood products.

Logging residue. hT e net volume of unused portions of the merchantable central stem of growing-stock trees cut or killed by logging.

Logging slash. The net volume of unused portions of the unmerchantable (non-growing stock) sections of trees cut or killed by logging.

Merchantable sections. Refers to sections of the central stem of growing-stock trees that meet either pulpwood or saw log specifications.

Net volume. Gross volume less deductions for rot, sweep, or other defects affecting use for roundwood products.

Noncommercial species. Trees species of typically small size, poor form, or inferior quality that normally do not develop into trees suitable for industrial roundwood products. Noncommercial species are listed in the volume tables as rough trees.

Nonforest land. Land that has never supported forests, and land formerly forested where use for timber management is precluded by development for other uses. (Note: Includes areas used for crops, active Christmas tree plantations, orchards, nurseries, improved pasture, residential areas, city parks, improved roads of any width and adjoining clearings, powerline clearings of any width, and 1- to 39.9-acre areas of water classified by the Bureau of the Census as land.) If intermingled in forest areas, unimproved roads and nonforest strips must be more than 120 feet wide and more than 1 acre to qualify as nonforest land.

Nonforest land removals. Net volume of trees on nonforest lands harvested for industrial roundwood products.

Poletimber. A growing-stock tree at least 5.0 inches d.b.h. but smaller than sawtimber size (9.0 inches d.b.h. for softwoods, 11.0 inches d.b.h. for hardwoods).

Poletimber removals. Net volume in the merchantable central stem of poletimber trees harvested for industrial roundwood products.

Primary wood-using mills. Mills receiving roundwood or chips from roundwood for processing into products such as lumber, veneer, and pulp.

Primary wood-using mill residue. Wood materials (coarse and fine) and bark generated at manufacturing plants that process industrial roundwood into principal products. These residues include wood products obtained incidental to production of principal products and wood materials not utilized for some product.

Production. The quantity of roundwood material harvested in a geographic area plus all roundwood material exported to other geographical areas.

Receipts. hT e quantity of roundwood material received by commercial mills in a geographic area plus all roundwood material imported from other geographical areas.

Retained. Roundwood volume harvested from and processed by mills within the same state.

Rotten tree. A tree that does not meet regional merchantability standards because of excessive unsound cull.

Rough tree. A tree that does not meet regional merchantability standards because of excessive sound cull (includes forks, sweep and crook, and large branches or knots), including noncommercial tree species.

Roundwood. Logs, bolts, or other round sections cut from trees (including chips from roundwood).

Sapling. A live tree between 1.0 and 5.0 inches d.b.h.

Saw log portion. That portion of the central stem of sawtimber trees between the stump and the saw log top.

Saw log top. The point on the central stem of sawtimber trees above which a saw log cannot be produced. The minimum saw log top is 7.0 inches diameter outside bark for softwoods and 9.0 inches diameter outside bark for hardwoods.

Sawtimber removals. As used in Table 10, sawtimber removals refers to the net volume in the merchantable central stem of sawtimber-size trees harvested for industrial roundwood products. (Note: includes the saw log and upper stem portions of sawtimber-size trees.) When referring to the sawtimber volume removed from timberland as in Table 12 and 12a, sawtimber removals refers to the net volume in the saw log portion of sawtimber-size trees harvested for roundwood products or left on the ground as harvest residue, and is usually expressed in thousands of board feet (International 1/4-inch rule or Scribner rule).

Sawtimber tree. A growing-stock tree containing at least a 12-foot saw log or two noncontiguous saw logs 8 feet or longer, and meeting regional specifications for freedom from defect. Softwoods must be at least 9.0 inches d.b.h. and hardwoods must be at least 11.0 inches d.b.h.

Sawtimber volume. Net volume in the saw log portion of sawtimber trees.

Scribner rule. A log rule or formula for estimating the board-foot volume of logs based on diagrams of perfect circles, with 1 inch thick boards of varying 2 inch multiple widths, positioned in the circle to provide the best utilization, and allowing for 1/4 inch of kerf. hT is rule is used in the Western United States.

Softwoods. Coniferous trees, usually evergreen, having needles or scale-like leaves.

Timber product output. The volume of roundwood products produced from an area's forests.

Timberland. Forest land that is producing, or is capable of producing, in excess of 20 cubic feet per acre per year of industrial roundwood products under natural conditions, is not withdrawn from timber utilization by statute or administrative regulation, and is not associated with urban or rural development.

Tree. A woody perennial plant, typically large, with a single well-defined stem carrying a more or less definite crown; sometimes defined as attaining a minimum diameter of 3 inches (7.6 cm) and a minimum height of 15 feet (4.6 m) at maturity. For FIA, any plant on the tree list in the current field manual is measured as a tree.

Upper stem portion. That portion of the central stem of sawtimber trees between the saw log top and the minimum top diameter of 4.0 inches outside bark, or to the point where the central stem breaks into limbs.

COMMON AND SCIENTIFIC NAMES OF TREE SPECIES BY SPECIES GROUP

Softwoods

Cedar/juniper
Eastern redcedar   *Juniperus virginiana*
Rocky Mountain juniper   *Juniperus scopulorum*
Spruce
White spurce   *Picea glauca*
Lodgepole pine   *Pinus contorta*
Ponderosa pine   *Pinus ponderosa*
Other Pines
Red pine   *Pinus resinosa*
White pine   *Pinus strobus*
Douglas-fir   *Pseudotsuga menziesii*

Hardwoods

Hard maple
Black maple   *Acer nigrum*
Sugar maple   *Acer saccharum*
Soft maple
Boxelder   *Acer negundo*

Red maple	*Acer rubrum*
Silver maple	*Acer saccharinum*
Hackberry	*Celtis occidentalis*
Ash	
Green ash	*Fraxinus pennsylvanica*
Honeylocust	*Gleditsia triacanthos*
Black walnut	*Juglans nigra*
Cottonwood	
Eastern cottonwood	*Populus deltoides*
Plains cottonwood	*Populus deltoides ssp. monilifera*
Aspen	
Quaking aspen	*Populus tremuloides*
Bur oak	*Quercus macrocarpa*
American basswood	*Tilia americana*
Elm	
American elm	*Ulmus americana*
Siberian elm	*Ulmus pumila*
Slippery elm	*Ulmus rubra*

REFERENCES

Choate, G.A.; Spencer, J., Jr. 1969. Forest in South Dakota. Resour. Bull. INT-8. Ogden, UT; U.S. Department of Agriculture, Forest Service, Intermountain Forest & Range Experiment Station. 40 p.

Collins, D.C.; Green, A.W. 1988. South Dakota's timber resources. Resour. Bull. INT56. Ogden, UT; U.S. Department of Agriculture, Forest Service, Intermountain Research Station. 48 p.

Hackett, R.L.; Sowers, R.A. 1996. South Dakota timber industry—an assessment of timber product output and use, 1993. Resour. Bull. NC-175. St. Paul, MN: U.S. Department of Agriculture, Forest Service, North Central Forest Experiment Station. 19 p.

Piva, R.J. 2010. South Dakota's forest resources, 2009. Res. Note. NRS-82. Newtown Square, PA: U.S. Department of Agriculture, Forest Service, Northern Research Station. 4 p. (Available only online at: http://www.nrs.fs.fed.us/pubs/rn/rn_nrs82.pdf).

Piva, R.J.; Josten, G.J. 2003. South Dakota timber industry—an assessment of timber product output and use, 1999. Resour. Bull. NC-213. St. Paul, MN:

U.S. Department of Agriculture, Forest Service, North Central Research Station. 31 p.

Piva, R.J.; Josten, G.J.; Mayko, Richard D. 2006. South Dakota timber industry—an assessment of timber product output and use, 2004. Resour. Bull. NC-264. St. Paul, MN: U.S. Department of Agriculture, Forest Service, North Central Research Station. 36 p.

U. S. Census Bureau. 2007. 2007 Economic Census. Available at: http://www. census.gov/econ/ census07/ (Accessed May 2, 2012).

INDEX